Access to Advanced Lev Chemistry

Series editors
Ted Lister and Janet Renshaw,
Trinity School, Leamington Spa

SIMON & SCHUSTER
EDUCATION

Contents

Acknowledgements

The authors thank those pupils at Trinity School who trialled the materials and also Pat Rowlinson at Simon & Schuster Education for her part in the production of this book.

The authors would like to thank the following for their kind permission to reproduce photographs: Educational Development Center 7.9, 7.12; GeoScience Features Picture Library 6.6; Claire Starkey 1.3, 4.1, 5.6, 6.3, 6.4 and 8.2.

Text © M Baker, J King, J Lister, J Renshaw

Design and artwork © Simon & Schuster 1992

First published in 1992 in Great Britain by
Simon and Schuster Education
Campus 400, Maylands Avenue
Hemel Hempstead, Herts HP2 7EZ

Reprinted in 1993

A catalogue record for this book is available from the British Library

ISBN 07501 02543

Typeset by Jim Weaver Design
Illustrated by Tek-Art
Printed in Great Britain by St Edmundsbury Press Ltd
Bury St Edmunds, Suffolk

Introduction

This book is designed to help you to get ready for a post-16 course in chemistry: Advanced Level, Advanced Supplementary Level, BTEC, Scottish Higher etc. The precise course you will be following doesn't matter because this book stresses the principles of chemistry which are the same for any course. You could use it before you begin your advanced course or during the first part of that course. It should also be useful for reference during your course – especially the chapters on essential physics and maths.* *Access to Advanced Level: Chemistry* has been designed so that you can work through it on your own and the answers to the questions are at the end of each chapter. However, cheating won't help your understanding!

How to use this book

Teaching yourself how to do something needs confidence, and often that is the one thing that you don't have. We suggest that you work through each section slowly and don't move on to the next section until you have correctly answered the questions. If you are getting most of them right, you are doing well. What if you aren't? One resource is a standard chemistry textbook. One of the skills you will have to develop for advanced study is independent learning. There are a variety of approaches to explaining concepts and ours may not always be the best for you. At advanced level (and beyond), referring to text books and reading on your own initiative is going to give you a valuable skill and that essential ingredient **confidence** in your ability to learn by yourself.

Good luck and enjoy your chemistry!

The authors and editors

*There is a glossary of important terms at the back of the book for you to refer to.

Chapter 1

The kinetic particle theory

Starting points

- You should be able to recall some simple experiments which suggest that matter is made of moving particles.

- Note 5×10^{-10} is another way of writing 0.0000000005. If you are not familiar with numbers expressed in this way, then refer to Chapter 8.

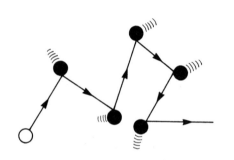

1.1 The path of a gas particle. It will make thousands of collisions every second.

Moving particles

One of the most basic ideas in science is that all substances are made up of tiny particles (atoms, molecules and ions). These particles are in constant motion and this leads to the name **kinetic particle theory**, the word kinetic meaning moving. A typical atom is about 5×10^{-10} m across and may be moving at speeds of 1000 km/hour. The three states of matter differ in the spacing, movement and arrangement of the particles as shown in Table 1. When a substance is heated, the particles move faster (gain kinetic energy) but do not change in size.

Table 1   Particles in solid, liquid and gas

	Solid	**Liquid**	**Gas**
Arrangement	regular	random	random
Spacing	touching	close	far apart
Movement	vibrating	rapid jostling	very rapid

Using the kinetic theory

The kinetic theory enables us to explain the properties of the three states of matter and a number of other things of importance to chemists.

Shape

The particles in a solid must be strongly held together so that solids have a definite shape. In liquids, the particles are slightly further apart and more loosely held so liquids fill the bottom of their containers. Particles in a gas are in rapid random motion so gases completely fill their containers.

Change of state

Particles in solids are vibrating about fixed points. If the solid is heated, the particles gain more energy and vibrate more. Eventually, the particles are vibrating so much that the regular arrangement breaks down and the particles become free to move around. The solid has melted and become liquid. Further heating transfers more energy

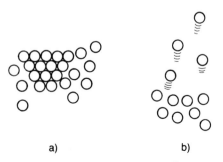

a)   b)

1.2   Pure substances a) melt b) boil at definite temperatures.

to the particles until at a particular temperature they become able to move completely freely. The spacing between the particles becomes very large indeed. The liquid has boiled – turned into a gas.

1 Use the kinetic theory and the information in Table 1 to explain why most solids expand when they are heated.

Gases

Gas pressure

All gases exert pressure on the walls of their containers. This is caused by the fast-moving gas particles which bombard the container walls. This pressure is not noticeable in everyday situations, because most containers have gas inside and out. This means that there is an equal and opposite bombardment on both sides and the two cancel out. However, if we use a pump to remove the air from inside an oil can, the effect of the pressure becomes obvious – see Fig 1.3.

1.3   An oil can before and after air has been pumped out of it.

Pressure and temperature

Gas pressure increases with temperature. At higher temperatures, the particles move faster and therefore hit the walls of the container both harder and more often. If we measure the pressure of a fixed mass of gas at different temperatures and plot a graph, we find that it is a straight line – see Fig 1.4.

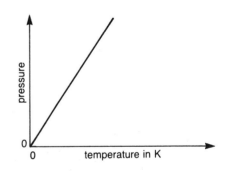

1.4   The straight line shows that pressure is directly proportional to temperature.

Pressure and volume

2 A syringe full of air is connected to a pressure gauge as shown in Fig 1.5.

What do you think will happen to the pressure if we squeeze the plunger in so that the gas volume is halved? Try to explain your answer in terms of what the particles of the gas are doing. Hint: how often do they hit the walls, and how hard do they hit the walls after squeezing compared with before?

1.5   A syringe of air connected to a pressure gauge.

1.6   Changing the temperature of the gas in the syringe will change its volume.

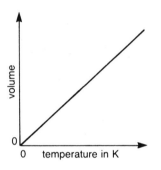

1.7   The straight line shows that the volume of a gas is directly proportional to the temperature. At 0 K the graph predicts that a gas would have no volume!

Volume and temperature

If a gas is kept in a container whose volume can change, such as a syringe (see Fig 1.6) – we find that changing the temperature affects the volume. Heating the gas makes the particles move faster so that they hit the walls of the container harder and more often. This pushes the plunger of the syringe which moves away, increasing the volume of the gas. If we measure the volume of a fixed mass of gas at different temperatures and plot a graph we get a straight line. See Fig 1.7.

The gas laws

The three quantities pressure, P, volume, V, and temperature, T, of a fixed mass of gas are linked mathematically. If we measure the temperature in degrees kelvin, then:

the volume is proportional to the temperature ($V \propto T$) if we keep the pressure constant. This is called **Charles' Law**.

the pressure is proportional to the temperature ($P \propto T$) if we keep the volume constant. This is sometimes called the **Constant Volume Law**.

the product $P \times V$ is always constant if we keep the temperature constant. This is called **Boyle's Law**.

Note   The **kelvin scale of temperature** has its zero at –273°C. We believe that this is the lowest possible temperature. It is often called absolute zero. At this temperature, molecules have no energy at all.

To convert °C to degrees kelvin (written K, not °K) we add 273.

> 3   a   Convert the following temperatures to K: 100 °C, 0 °C, 500 °C.
>   b   Convert the following temperatures to °C: 500 K, 473 K, 200 K.

Reaction rates

Chemical reactions can only happen if the reacting particles collide with one another. The more often and harder they collide, the faster the reaction will be. This means that the following factors will increase reaction rates:

a   increasing the temperature

b   increasing the the surface area of a solid reactant

c   increasing the pressure of a gas

d   increasing the concentration of a solution.

> 4   Explain how each of the above factors increases the rate of a reaction in terms of what the particles are doing.

Diffusion

Diffusion is the word used to describe the fact that gases and liquids tend to mix of their own accord (ie without draughts or stirring). For example we can soon tell (by smelling) if the gas has been left on in the kitchen. Diffusion happens because particles are moving fast and there are gaps between the particles in both the solid and the liquid state.

1.8 Rotten smells.

5 Diffusion takes place much more quickly in gases than in liquids. For example the smell of a bad egg reaches your nose almost instantly while a drop of ink will take many minutes to spread throughout a beaker of water. Explain this difference in terms of what the particles are doing and how they are arranged.

Entropy

An interesting point is that while gases mix completely and readily of their own accord, they will not unmix. We never have to worry about the oxygen and nitrogen in the air unmixing so that all the oxygen goes over to one side of the room and all the nitrogen to the other! This simple observation has a lot to tell us about why some chemical changes happen and others do not. Changes which involve going from order (like two separate gases) to disorder (like a mixture of gases) tend to happen of their own accord. This is rather like the fact that it is easy to mix up a pack of cards but much more effort to arrange them in order. The degree of disorder in a chemical system is called its **entropy**.

1.9   The separate gases have more order than the mixture, so gases mix of their own accord.

Answers

1   The particles gain extra energy which makes them vibrate more thus pushing one another further apart.

2   The pressure will increase (double). The particles do not change in speed so they hit the walls of the syringe just as hard as before. As the volume is now smaller, they hit the walls more often.

3   a   373 K; 273 K; 773 K
    b   227 °C; 200 °C; –73 °C

4   a   The particles of reactants will collide with one another harder and more often.
    b   The particles of the other reactant will find more exposed particles of the solid reactant to collide with.
    c   The reactant particles will collide more often as there is less space in which to move.
    d   More particles of reactant in the same space will lead to more collisions.

5   In a gas, the particles are moving faster and there is more space between them.

Chapter 2

Atomic structure and bonding

Starting points

- Do you understand the following terms: **element, compound, atom, relative atomic mass**? Write a sentence or two (no more) to explain what you mean by each term. Then check the glossary at the back of the book.

- Expressing numbers as powers of 10. Do you know that 1×10^{-3} is another way of writing 0.001? If you are not familiar with this, see Chapter 8.

	flea	grain of sand	red blood cell
⊔⊔⊔⊔⊔⊔ 1×10^{-2}	(flea) 1×10^{-3}	◈ 1×10^{-4}	◎ 1×10^{-5}

bacteria	virus	haemo-globin molecule	sugar molecule
⌁ⵔ 1×10^{-6}	⬡ 1×10^{-7}	1×10^{-8}	1×10^{-9}

atom			
1×10^{-10}			

proton or neutron	approximate size in metres
1×10^{-14}	

2.1 An idea of the relative size of the atom.

A closer look at atoms

Sub-atomic particles

Scientists believe that atoms are made up of three main sub-atomic particles, the **proton**, the **neutron** and the **electron**. Their properties are summarised in Table 2.1.

Table 2.1 Sub-atomic particles

	proton	neutron	electron
approximate mass in u *	1	1	1/1840 (approximately 0)
electric charge	+1	0	−1

* 1 u is approximately 1.6×10^{-27} kg

The information in the **Periodic Table** enables us to find the number of protons, neutrons, and electrons in any atom. Since the electrons have practically no mass, the mass of the atom is simply the sum of the masses of the protons and neutrons.

The **atomic number** is the number of protons in an atom. It tells us what element we have because the number of protons is the same as the number of electrons and the electrons are responsible for the chemical reactions of an element. For example, in sodium, Na:

$23 \leftarrow$ relative atomic mass, A_r (= protons + neutrons)
Na
$11 \leftarrow$ atomic number, Z (= protons = electrons)

So in an atom of sodium, Na :

number of protons   = 11
number of electrons = 11
number of neutrons = 23 − 11 = 12

1 Work out the numbers of protons, neutrons and electrons in the following atoms: carbon, C; lithium, Li; uranium, U; manganese, Mn.

2 a What is unique about the make-up of the sub-atomic particles in a hydrogen atom, H?
   b What seems strange about the make-up of the sub-atomic particles in a chlorine atom, Cl?

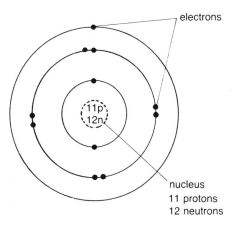

2.2   The sub-atomic particles in a sodium atom.

Arrangement of sub-atomic particles

These sub-atomic particles are arranged as shown in Fig 2.2 to make up the atom. The protons and neutrons are tucked away in the centre of the atom called the **nucleus**, held together by the strong nuclear force. The electrons orbit the nucleus in a series of orbits or shells of increasing distance from the nucleus.

The electrons are held to the nucleus by the attraction of opposite electric charges but they do not fall into it because of their orbital motion.

The scale of the atom

A typical atom has a nucleus which is about 1×10^{-14} m across while the atom itself is about 1×10^{-10} m across. This makes the nucleus some ten thousand times smaller than the whole atom. This would make an atom the size of a soccer pitch have a nucleus the size of a marble.

Isotopes, the strange case of chlorine

Look back at your answer to question 2b. There are not really any $\frac{1}{2}$ neutrons in the nucleus of a chlorine atom. The relative atomic mass shown in the Periodic Table is an average of two sorts of chlorine atom that exist. These are identified by their relative atomic masses as ^{35}Cl and ^{37}Cl. Both types of atom have the same atomic number, 17. They differ in that ^{35}Cl has 18 neutrons while ^{37}Cl has 20. ^{35}Cl atoms are three times as common as ^{37}Cl atoms. This means that the total relative atomic mass of four chlorine atoms picked at random would be

$(3 \times 35) + (1 \times 37) = 105 + 37 = 142$

This gives the average relative atomic mass of each atom as $142/4 = 35.5$

Atoms of the same element which have different numbers of neutrons are called **isotopes**.

- In the Periodic Table in this book (and many others) most of the relative atomic masses have been rounded to the nearest whole number.

3 What would be the numbers of protons, neutrons and electrons, in the following isotopes?
   a) $^{12}_{6}$C, $^{13}_{6}$C, $^{14}_{6}$C   b) $^{1}_{1}$H, $^{2}_{1}$H, $^{3}_{1}$H

4 Neon, Ne, has isotopes $^{20}_{10}$Ne and $^{22}_{10}$Ne. They exist in the approximate ratio 9:1. What will be the average relative atomic mass of neon?

Arrangement of electrons

The electron shells can hold different numbers of electrons. In order of increasing distance from the nucleus, these are:

First shell   up to 2
Second shell   up to 8
Third shell   up to 8 with reserve space for 10 more

The first shell is always filled before the second and so on.

We can draw electron arrangements as in Fig 2.2 or use shorthand, for example sodium: Na, 2,8,1.

5 Draw or write the electron arrangements for   a boron, B; b fluorine, F;   c lithium, Li.

Uncovering the structure of the atom

The full story of this would take up too much room for this book but it is a fascinating tale and well worth looking up. Below is a list of some of the main events and dates. The story is, of course, still going on.

1803 Dalton's model of the atom as a small billiard ball (the idea used in kinetic theory).
Late 19th century Discovery of radioactive elements (their break-up reveals that atoms have structure).
1897 Joseph John ('J J') Thomson discovers the electron (a negatively charged constituent of all atoms).
1890s J J Thomson's model of the atom as a positive sphere embedded with rings of electrons (the currant bun model).
1909 Geiger, Marsden and Rutherford's alpha scattering experiment shows that atoms have a nucleus. (See Fig 2.3)
1911 Rutherford's model of the atom has electrons orbiting the nucleus.
1913 Bohr suggests electron shells.
1920 Rutherford names the proton.
1920s Electrons considered as clouds of charge rather than as particles.
1932 Chadwick discovers the neutron.
To the present day Discovery of even more sub-atomic particles such as **muons, quarks, charmed particles** and so on. You might like to look some of these up.

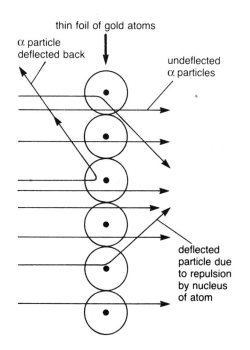

2.3   Most of the α particles passed through the thin foil but a few were deflected back by collision with the nuclei of the gold atoms.

Why do atoms react?

Look at the three atoms in Fig 2.4.

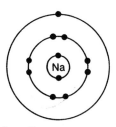

2.4   The electron arrangements of fluorine F, neon, Ne and sodium, Na.

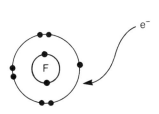

2.5 A fluorine atom gains an electron to get a full outer shell.

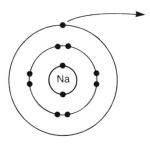

2.6 A sodium atom loses an electron (and its outer shell) to leave it with full electron shells.

The way in which they react is linked to their electron arrangements. Neon is unreactive. It has full shells of electrons.

Fluorine always reacts to get an extra electron. This gives it full electron shells (Fig 2.5).

Sodium always reacts to lose one electron. This gives it full electron shells (Fig 2.6).

When they react as shown, both fluorine and sodium become **ions**.

Ions

An ion is a charged atom (or group of atoms). Atoms that gain extra electrons become **negative ions**.

For example a fluorine atom (9 protons and 9 electrons) becomes a fluor*ide* ion, F^-, (9 protons and 10 electrons) and a sulphur atom (16 protons and 16 electrons becomes a sulphide ion, S^{2-}, (16 protons and 18 electrons). **Note** negative ions change the ending of their names to -ide.

Atoms that lose electrons become **positive ions**. For example a sodium atom (11 protons and 11 electrons) becomes a sodium ion, Na^+, (11 protons and 10 electrons) and a magnesium atom (12 protons and 12 electrons) becomes a magnesium ion, Mg^{2+}, (12 protons and 10 electrons).

6 Neon atoms have 10 electrons (2,8) as do F^- ions and Na^+ ions. Why don't fluorine and sodium *become* Ne when they react?

7 Give the charge of the ions you would expect to be formed by a calcium, b oxygen, c aluminium d nitrogen.

Electron arrangements of the first twenty elements

I	II	III	IV	V	VI	VII	0
H 1							**He** 2
Li 2, 1	**Be** 2, 2	**B** 2, 3	**C** 2, 4	**N** 2, 5	**O** 2, 6	**F** 2, 7	**Ne** 2, 8
Na 2, 8, 1	**Mg** 2, 8, 2	**Al** 2, 8, 3	**Si** 2, 8, 4	**P** 2, 8, 5	**S** 2, 8, 6	**Cl** 2, 8, 7	**Ar** 2, 8, 8
K 2, 8, 8, 1	**Ca** 2, 8, 8, 2						

Notice that the number of electrons in the outer shell is the same as the group number in the Periodic Table, except for Group 0 (which you could think of as having no electrons following a full outer shell).

Bonding

- All electrons are identical, but drawing some as dots and some as crosses helps us to see what is going on.

Where do all the electrons go to or come from? The answer is that atoms do not react singly, they need other atoms to react or bond with. We can keep track of what happens to electrons using **dot-cross diagrams**.

Metal and non-metal

These bond together **ionically**.

Sodium fluoride

Na (2,8,1)
(11p, 11e⁻)

sodium atom

F (2,7)
(9p, 9e⁻)

fluorine atom

REACTION

Na⁺ (2,8)
(11p, 10e⁻)

sodium ion

F⁻ (2,8)
(9p, 10e⁻)

fluoride ion

sodium fluoride NaF

2.7   The formation of sodium fluoride.

○ Fluoride ion, F⁻

● Sodium ion, Na⁺

2.8   The sodium and fluoride ions are regularly arranged in a giant structure.

Sodium needs to give away one electron to get a full outer shell and fluorine needs to gain one, so the formula is NaF (one to one).

Although the dot-cross diagram shows just one atom of each element, the ionic bonding extends throughout the structure. All the ions are held together in a **giant structure**. The bonding is strong and so the compound is hard to melt.

Magnesium fluoride

Magnesium needs to give away two electrons to get a full outer shell but fluorine needs to gain only one. So two fluorines are needed for each magnesium and the formula is MgF₂ (one to two). Magnesium fluoride also forms a giant structure.

Mg (2,8,2)

F (2,7)

F (2,7)

REACTION

Mg²⁺ (2,8)

2 × F⁻ (2,8)

MgF₂ magnesium fluoride

2.9   Dot-cross diagrams to show how magnesium fluoride is formed.

> 8 Draw similar diagrams to show the bonding between atoms of:
>   a sodium and chlorine
>   b sodium and oxygen
>   c calcium and fluorine
>
> 9 What is the formula and name of each of the products in question 8?

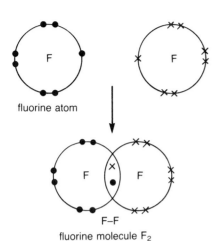

fluorine atom

F–F
fluorine molecule F₂

2.10   The bonding in fluorine gas F_2.

Properties of compounds with ionic bonding

- They are solids with high melting points.
- Ionic solids cannot conduct electricity as the ions are held fast.
- Melted ionic compounds can conduct electricity as the ions are now free to move.
- Many ionic compounds are soluble in water.
- If the compounds dissolve in water the solutions conduct electricity as the ions are now free to move.

Non-metal and non-metal

All non-metal atoms have (except for the noble gases) gaps in their outer shells, but in this case there are no metals to provide electrons to fill the gaps. Bonding between two non-metal atoms occurs by sharing of electrons. This is called **covalent** bonding. Small groups of atoms joined by covalent bonding are called **molecules**.

Two fluorine atoms

Each fluorine atom is bonded to just one other fluorine *within* the molecule. (See Fig 2.10.) In a sample of fluorine gas there would, of course, be vast numbers of other F₂ molecules, each one independent of the others. This molecular structure contrasts with the giant structures we have seen formed by ionic bonding.

Notice that:

- there is no transfer of electrons.
- there are no charged particles.
- the covalent bond is between only the atoms sharing the pair of electrons.
- there is no strong attraction *between* molecules.

Water (Fig 2.11) and oxygen (Fig 2.12)

Notice that the water molecule has two covalent bonds.

Notice that a total of four electrons is shared between the two oxygen atoms. This is called a double bond.

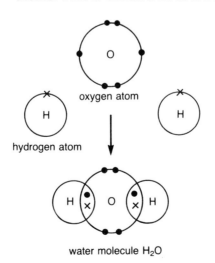

oxygen atom

hydrogen atom

water molecule H₂O

2.11   The bonding in water.

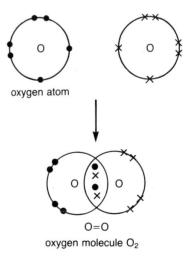

oxygen atom

O=O
oxygen molecule O₂

2.12   Oxygen molecules have a double bond.

10   Draw dot-cross diagrams to show the bonding in:
a   Chlorine, Cl_2
b   Methane, CH_4
c   Nitrogen, N_2

Giant covalent structures

As well as small groups of atoms called molecules, covalent bonding can produce giant structures. Such structures have a network of covalent bonds throughout the compound. This means that a lot of bonds have to be broken to melt the compounds and so they have high melting points. (See Fig 2.13.)

Properties of covalently bonded compounds

Covalently bonded compounds, both molecular and giant:
- are poor conductors of electricity as there are no charged particles.
- often do not dissolve in water.
- if they do dissolve in water, their solutions are poor conductors of electricity.

Those with molecular structures are gases, liquids or low melting point solids. Those with giant structures are high melting point solids.

- silicon atom   — represents a covalent bond, a pair of shared electrons
- oxygen atom

2.13   Silicon dioxide (the main component of sand) has a giant covalent structure.

Metals

All metal atoms have one, two or three electrons in their outer shells. They can bond to other metal atoms by contributing these electrons to a general pool of electrons that are free to move throughout the resulting giant structure. See Fig 2.14.

It is the mobile electrons that give metals and alloys their unique properties.

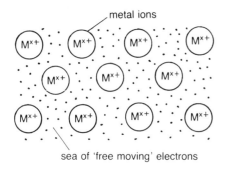

metal ions

sea of 'free moving' electrons

2.14   The bonding in a metal.

Alloys – 'designer metals'

Mixtures of different metals are called alloys. Alloys can be 'tailored' to match their properties to particular needs. The human race uses a large number of alloys for different purposes. Some examples include:

solder – tin and lead – for low melting point (for a metal)
brass – copper and zinc – easy to work
bronze – copper and tin – hard and corrosion resistant
duralumin – aluminium and copper – strong and light

Properties of substances with metallic bonding

- They have structures with high melting points.
- They conduct electricity as solids or liquids.
- They are shiny, malleable (can be shaped by beating) and ductile (can be pulled into wires).

11 Which is the only metal element that does not have a giant structure at room temperature?

Structure and bonding – Summary

BONDING describes what holds atoms together. There are three types which occur in different situations and result in different properties.

Covalent bonding occurs by electron-sharing between non-metal atoms. The resulting compounds do not conduct electricity in any physical state.

Ionic bonding occurs by electron transfer from a metal to a non-metal. The resulting compounds conduct electricity when liquid or in solution but not when solid.

Metallic bonding occurs by pooling of electrons between metal atoms. Metals conduct both as liquids and solids.

STRUCTURE describes the arrangement of atoms. There are two main types – giant and molecular.

Giant structures have a regular 3-D arrangement of atoms which extends throughout the compound. The structure may be held together by any of the three types of bonding. Giant structures have high melting and boiling points.

Molecular structures consist of many small units (molecules) which each consist of a few atoms held together by covalent bonds. The molecules are not strongly bonded to one another although the bonding *within* them is strong. Molecular structures can be recognised by their low melting and boiling points.

12 The Table 2.2 below gives some information about four substances. State which one is ionic, which metallic, which has a molecular structure and which has a covalent giant structure.

Table 2.2

Substance	Melting point °C	Boiling point °C	Electrical conduction	
			As solid	As liquid
A	1083	2567	good	good
B	–182	–164	poor	poor
C	1723	2230	poor	good
D	993	1695	poor	poor

Some glimpses ahead

X-ray diffraction

The method of X-ray diffraction provides direct evidence for some of our theories of bonding. This technique allows us to see where the electrons are in a molecule or giant structure. The electrons show up rather like contour lines on a map.

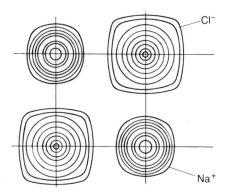

2.15   An electron density map of sodium chloride, NaCl, an ionic compound.

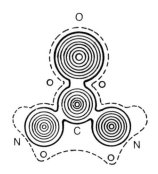

2.16   An electron density map for urea, $CO(NH_2)_2$, a covalent compound.

Fig 2.15 represents an ionic compound as the separate ions are obvious while Fig 2.16 is a covalent compound as the sharing of electrons can be seen.

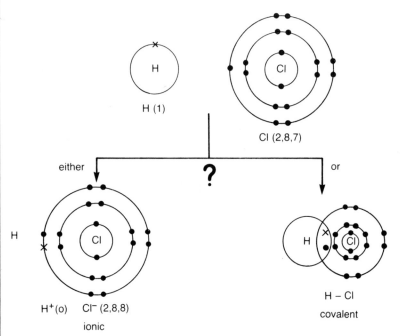

2.17   Hydrogen and chlorine could be bonded covalently or ionically.

The bonding of HCl

The dot-cross diagrams in Fig 2.17 show two ways in which an atom of hydrogen could bond to one of chlorine – one ionic and one covalent. Which is correct? The answer seems to be both.

Pure hydrogen chloride is a gas which does not conduct electricity – just what we would expect for a covalently bonded compound. However, when we dissolve it in water, hydrogen chloride does conduct electricity – as we would expect for an ionic compound. The truth seems to be that the bonding in this substance (and many others) is somewhere between the two extremes.

Answers

1
	p	n	e
C	6	6	6
Li	3	4	3
U	92	146	92
Mn	25	30	25

2 a Hydrogen atoms have no neutron.
  b Chlorine appears to have 18.5 neutrons.

3 a
	p	n	e
$^{12}_{6}C$	6	6	6
$^{13}_{6}C$	6	7	6
$^{14}_{6}C$	6	8	6
b $^{1}_{1}H$	1	0	1
$^{2}_{1}H$	1	1	1
$^{3}_{1}H$	1	2	1

4 20.2

5 a 2,3; b 2,7; c 2,1

6 Only the number of electrons changes. The nucleus, which gives the atom its identity, does not vary.

7 a Ca^{2+}, b O^{2-}, c Al^{3+}, d N^{3-}

8 a

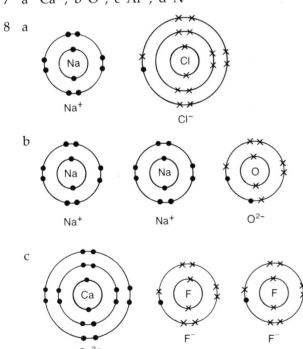

9 a NaCl, sodium chloride; b Na_2O, sodium oxide;
  c CaF_2, calcium fluoride

10 a

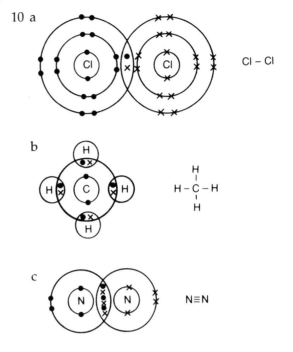

11 Mercury, Hg, a liquid!

12 A metallic; B molecular; C ionic;
   D giant covalent

Chapter 3   Formulae

Starting points

- Do you understand the following terms: **atom, element, compound, molecule, ion, giant structure, molecular structure, covalent bonding**? Write a sentence or two (no more) to explain what you understand by each term then check with the glossary.

- Note Chemical symbols normally have one or two letters. In two-letter symbols, the first letter is always a capital and the second letter small. So Co is the symbol for cobalt, while CO stands for one atom of carbon and one of oxygen. You will need to use the Periodic Table (inside back cover) to look up the element that each symbol represents.

The formula of a compound is useful because it indicates how many particles of each element are present in one **entity** of the compound.

The word entity is used to describe the simplest formula unit of a compound. It can be used to refer to compounds with molecular structures such as water, H_2O, or ammonia, NH_3 or giant ionic such as sodium chloride, NaCl, or calcium chloride, $CaCl_2$.

CaS means one atom of calcium to one atom of sulphur.
$CaCl_2$ means one atom of calcium to two atoms of chlorine.
H_2SO_4 means two atoms of hydrogen to one atom of sulphur to four atoms of oxygen

1 Say how many atoms of which elements are combined together in:

  a KF   b NaCl   c C_2H_6 d $C_6H_5NO_2$

Brackets

Sometimes formulae are written using brackets. A small number after a bracket multiplies all the elements in the bracket by that number.

$Ca(OH)_2$: each entity contains 1 calcium, 2 oxygen and 2 hydrogen atoms
$Ca(NO_3)_2$: each entity contains 1 calcium, 2 nitrogen and 6 oxygen atoms

2 Say how many atoms of each element are present in:

  a $(NH_4)_2SO_4$ b $Mg(NO_3)_2$ c $Ca_3(PO_4)_2$

Naming compounds

When the compound consists of a metal and a non-metal the naming is straightforward, we put the names of the two elements together, metal first, and change the ending of the non-metal to **-ide**. This ending signifies that there are only two elements in the compound.

	(METAL NAME)	(NON-METAL NAME) ('IDE')
NaCl	Sodium	chloride
KI	Potassium	iodide
$CaCl_2$	Calcium	chloride

3 Name the following:
   a LiBr   b Na_2O c MgO
   d MgI_2 e FeS   f Al_2O_3

Sometimes part of the compound consists of a group of particles. For example, the group SO_4. Some groups of atoms are given special names. Ones you will often meet are

SO_4 sulphate
NO_3 nitrate
CO_3 carbonate   } These behave as non-metals
PO_4 phosphate   and form the second part of the name.
OH   hydroxide
NH_4 ammonium   This behaves as a metal and forms the first
                                     part of the name.

From these examples you may be able to see that the ending **-ate** is often used to mean 'and oxygen as well' although hydroxide is an exception.

4 Name the following
   a $Cu(NO_3)_2$ b $Ca(OH)_2$ c $AlPO_4$
   d $MgSO_4$ e Li_2CO_3 f NH_4Cl

Formulae and bonding

Why do compounds have constant formulae? Why is water always H_2O rather than HO or HO_2? Why is sodium chloride always NaCl rather than Na_2Cl or $NaCl_2$? The reason is found in the way the atoms in the compound are bonded (see Chapter 2).

Ionic compounds

These contain a metal and a non-metal. Metals always form positive ions and non-metals negative ions. Table 3.1 gives some examples.

3.1   Which compound named in question 4 might be relevant to this policeman?

Table 3.1   The charges on some ions

Positive ions			Negative ions		
1+	2+	3+	1-	2-	3-
Li^+ lithium	Mg^{2+} magnesium	Al^{3+} aluminium	Cl^- chloride	O^{2-} oxide	PO_4^{3-} phosphate
Na^+ sodium	Ca^{2+} calcium		Br^- bromide	S^{2-} sulphide	
K^+ potassium	Cu^{2+} copper		I^- iodide	CO_3^{2-} carbonate	
NH_4^+ ammonium	Zn^{2+} zinc		NO_3^- nitrate	SO_4^{2-} sulphate	
			OH^- hydroxide		

Chapter 2 explains how we can predict the charges on some of these ions.

Since all compounds are neutral, we can predict the formula of an ionic compound. The total positive charge must be the same as the total negative charge.

Copper sulphate consists of two types of ion, copper Cu^{2+} and sulphate SO_4^{2-}. The compound resulting from these particles must be neutral overall and each Cu^{2+} is balanced by one SO_4^{2-}. So the formula is $CuSO_4$. In aluminium oxide, aluminium forms Al^{3+} and oxygen O^{2-}. The simplest way to get a neutral compound is if two aluminium ions (total charge 6+) combine with three oxide ions (total charge 6-), so the formula is Al_2O_3.

> 5 Use Table 3.1 above to predict the formulae of the following compounds:
>
> | | | | |
> |---|--------------------|---|-----------------------|
> | a | lithium carbonate | e | ammonium phosphate |
> | b | aluminium iodide | f | calcium oxide |
> | c | sodium nitrate | g | copper chloride |
> | d | zinc sulphate | h | aluminium phosphate |

Metal ions

- Metal ions are always positive.

- Metals in groups I, II and III have the same charge as their group number when they form ions.

- Transition metal atoms can often form two or more ions with different charges. Some common transition metal ions are given in Table 3.2.

Table 3.2   Some transition metal ions

1+	2+	3+
Ag+	Cu²⁺	Cr³⁺
Au+	Mn²⁺	Fe³⁺
Cu+	Fe²⁺	

Non-metal ions

- Non-metal elements in group VII always have a charge of 1- when they bond ionically.

- Non-metal elements in group VI always have a charge of 2- when they bond ionically.

- Some non-metal ions consist of more than one atom. The charges of some of these complex non-metal ions are given in Table 3.1.

Covalently bonded compounds

These are bonded together by electron-sharing. How many other atoms a particular atom can bond with depends on how many electrons it needs to gain a full outer shell, see Chapter 2.

Oxygen with six electrons in its outer shell needs two more, so it could bond with two hydrogen atoms which each need one extra electron to form water, H_2O. Each oxygen forms two **single bonds** (two electrons shared). See Fig. 3.2.

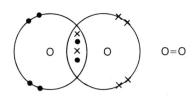

3.2 The bonding in water, H_2O. Oxygen forms two single covalent bonds.

3.3 Oxygen can form two covalent bonds to make a double bond.

Alternatively it could bond with one other oxygen so that both atoms share a total of four electrons. This is a **double bond.** See Fig. 3.3.

Combining power (valency)

Because it can form only one single bond, hydrogen is said to have a combining power or **valency** of one. Oxygen, which can form either two single bonds or one double bond, has a combining power of two.

6 Use the Periodic Table to help you work out the electron arrangement of the following atoms and use it to predict the combining power of each.

a Nitrogen, N
b Carbon, C
c Chlorine, Cl

d Sulphur, S
e Argon, Ar
f Silicon, Si

What do you notice about the combining power of carbon and silicon which are in the same group?

Some formulae

Here are the formulae of some interesting compounds.

aspirin $C_9H_8O_4$

TNT $C_7H_5N_3O_6$

insulin (the hormone injected by diabetics to control their blood sugar levels) $C_{254}H_{377}N_{65}O_{75}S_6$

ethyl mercaptan (according to the Guinness Book of Records, the most evil-smelling compound known – its smell is a cross between rotting cabbage, garlic, onions and sewer gas) C_2H_5SH

ethanol (the alcohol in alcoholic drinks) C_2H_5OH

Notice how similar the formulae of the last two are – what a difference an atom makes!

3.4 'My mouth tastes like C_2H_5SH, I've had too much C_2H_5OH – I need some $C_9H_8O_4$.'

Shapes of molecules

Formulae can tell us much more than just the numbers of different types of atoms in a molecule. They can tell us which atoms are bonded to which and even the shape of the molecule. Shape has a great deal to do with how molecules like drugs, for example, work. Chemists can now tailor compounds to the shape they want – designer molecules! They need to use computer graphics like the ones below to record the formulae.

Key
● carbon
◉ oxygen
○ hydrogen

3.6   Buckminsterfullerene – a molecular football, C_{60}, which consists entirely of carbon atoms. Drawing it in this way is called skeletal notation and one carbon atom is implied at every junction.

3.5   Computer graphic of aspirin, one of the oldest drugs known, but it is still finding new uses.

Bonding within the nitrate ion

The nitrate ion, NO_3^-, is an ion but it consists of four atoms covalently bonded together. Here are some facts about it:

- it is composed of one nitrogen and three oxygens
- nitrogen and oxygen are both non-metals and therefore the bonding within the ion is covalent
- the arrangement of the particles is such that nitrogen is surrounded by the three oxygens
- nitrogen has five electrons in its outer shell and oxygen has six.

Fig 3.7 shows a dot-cross diagram (see Chapter 2) for the covalent bonding within the nitrate ion.

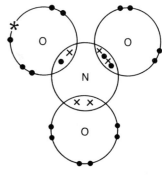

3.7   The nitrate ion, NO_3^-.

Notice
- All the atoms now have full outer electron shells.
- One oxygen has formed a double bond with the nitrogen (symbol =).
- One oxygen has formed a covalent bond with the nitrogen in which both the shared electrons came originally from the nitrogen atom. This is called a dative bond (symbol →, showing which atom has donated the pair of electrons to the other).
- If you count the electrons, you will find that there is one more than we started with. This has come from the metal atom which goes with the nitrate ion. This electron is marked * in the diagram. One extra electron is needed to allow each atom to have a full outer shell of electrons.

7 Draw dot-cross diagrams like the one in Fig 3.7 for the nitrate ion for the following ions:

a the hydroxide ion OH⁻.

b the carbonate ion CO_3^{2-}.
This is similar in structure to the nitrate ion but carbon has only four outer shell electrons.

Answers

1 a One atom of potassium to one atom of fluorine
 b One atom of sodium to one atom of chlorine
 c Two atoms of carbon to six atoms of hydrogen
 d Six atoms of carbon to five atoms of hydrogen to one atom of nitrogen to two atoms of oxygen.

2 a Two atoms of nitrogen to eight atoms of hydrogen to one atom of sulphur to four atoms of oxygen
 b One atom of magnesium to two atoms of nitrogen to six atoms of oxygen
 c Three atoms of calcium to two atoms of phosphorus to eight atoms of oxygen

3 a lithium bromide; b sodium oxide; c magnesium oxide;
 d magnesium iodide; e iron sulphide; f aluminium oxide

4 a copper nitrate; b calcium hydroxide; c aluminium phosphate;
 d magnesium sulphate; e lithium carbonate;
 f ammonium chloride

5 a Li_2CO_3; b AlI_3; c $NaNO_3$; d $ZnSO_4$; e $(NH_4)_3PO_4$;
 f CaO; g $CuCl_2$; c $AlPO_4$

6 a 2,5; three; b 2,4; four; c 2,8,7; one; d 2,8,6; two; e 2,8,8; zero;
 f 2,8,4; four

7 a

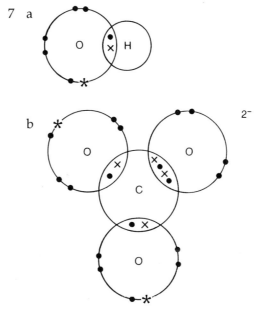

b

Chapter 4   Equations

Starting points

- You should know that equations represent the reactions that happen in real life. They are found by experiment. This means that just because you can write an equation, it doesn't follow that a reaction will actually happen.

- You should recognise the chemical symbols for the most commonly used elements and know how to look up others in the Periodic Table.

1 You should be able to do the following.

   a Write a word equation for what happens when magnesium is burned in oxygen.

   b Say what the following state symbols mean in an equation – (s) , (l), (g), (aq)

4.1   Magnesium burning in air. A little of the magnesium will also combine with the nitrogen in air.

Word and symbol equations

Word equations tell us which chemicals will react together and what products they form. Balanced symbol equations have a different function. They tell us in what proportions substances react together.

2 Can you recognise a **balanced equation**? Which of the following equations is/are balanced?

a	$2 H_2$	+	O_2	→	$2 H_2O$	
b	CH_4	+	O_2	→	CO_2	+	H_2O
c	Na	+	H_2O	→	$2 NaOH$	+	H_2
d	H_2	+	Cl_2	→	$2 HCl$	

A balanced equation has the same number of each type of atom on each side of the arrow, because chemical reactions simply rearrange atoms, they do not create or destroy them.

Although equations can only be found for certain by experiment, we can often work out what the balanced equation for a given reaction must be if we know the reactants (starting materials), the products and their formulae.

3 Balance the two equations below. (Don't change the formulae!)

a	Li	+	F_2	→	LiF
b	Fe	+	O_2	→	Fe_2O_3

Check your answers to question 3 in the answer section. If you were right, and feel confident about balancing equations, move on to the **Types of reactions** section. Otherwise read the section that follows on balancing equations.

Balancing equations – introduction

Suppose we want to balance the equation
$$Na \quad + \quad Cl_2 \quad \rightarrow \quad NaCl$$

The problem with the equation above is that there are two chlorine atoms on the left-hand side and only one on the right-hand side.

Step 1 To get two chlorines on the right-hand side put a 2 in front of NaCl

$$Na \quad + \quad Cl_2 \quad \rightarrow \quad 2\,NaCl$$

But now the chlorine is correct and not the sodium.

Step 2. Go back to the left-hand side and put a 2 in front of the Na.

$$2\,Na \quad + \quad Cl_2 \quad \rightarrow \quad 2\,NaCl \qquad -\ balanced\ -$$

Balancing equations

Balancing an equation is a bit like the problem of buying batteries for a torch. If the batteries come in packs of two and the torches take three batteries each, what is the smallest number of items to buy to get a set of working torches with nothing left over? A little thought should convince you that if you buy three packs of batteries and two torches you can assemble two working torches.

torch + batteries → working torch
2T + 3B$_2$ → 2TB$_3$
(where B = 1 battery, T = 1 torch)

4.2

This is just like balancing the equation for the reaction of aluminium (Al) with chlorine (Cl) to give aluminium chloride ($AlCl_3$).

$$2\,Al \quad + \quad 3\,Cl_2 \quad \rightarrow \quad 2\,AlCl_3$$

Hints for balancing equations

Remember that a balanced equation must have the same number of atoms of each element on each side of the equation. Here are a few hints.

- You must not change the formulae – these have been found by experiment and cannot be changed at your convenience.
- You can only change the number of atoms by putting numbers *in front of* the formulae.
- You often need more than one step to get it right.
- If you can't balance the equation in 3-4 steps, you probably need to check that you have all the formulae correct.

4   Balance the following equations

a	Mg	+	HCl	→	MgCl₂	+	H₂
b	Na	+	O₂	→	Na₂O		
c	Ca(OH)₂	+	HNO₃	→	Ca(NO₃)₂	+	H₂O
d	Ca	+	H₂O	→	Ca(OH)₂	+	H₂

a   Mg + HCl → $MgCl_2$ + H_2
b   Na + O_2 → Na_2O
c $Ca(OH)_2$ + HNO_3 → $Ca(NO_3)_2$ + H_2O
d   Ca + H_2O → $Ca(OH)_2$ + H_2

Types of reactions

Atoms in chemical substances can rearrange in all sorts of ways when they react but there are groups of reactions which follow the same patterns and which have their own names.

Direct combination

Writing the equation for two elements reacting together is probably the easiest one to do. As you have seen in the the chapter on formulae, the name of a compound of two elements always ends in -ide, so the word equation is straightforward. The difficult part is getting the formulae of the resulting compounds correct. You may be able to look them up or work them out – see Chapter 3.

4.3   Did you know that popping hydrogen is a direct combination reaction?

5   Write the word equation and then the balanced symbol equations for the following:

    a   iron reacting with sulphur
    b   magnesium reacting with nitrogen
    c   sodium reacting with bromine

(The formulae you need to use are NaBr, Mg_3N_2, FeS).

Oxidation and reduction

Oxidation

This, at its simplest, is what happens when we burn substances in air thus adding oxygen to them.

a   Metals and non-metals oxidise to their oxides and the most difficult part of these equations is knowing the formula of the oxide.

$$copper + oxygen \rightarrow copper\ oxide$$
$$2\,Cu + O_2 \rightarrow 2\,CuO$$

b   Or we can have compounds which already contain oxygen reacting to get more.

$$2\,SO_2 + O_2 \rightarrow 2\,SO_3$$

Sulphur dioxide has been oxidised to sulphur trioxide.

6 Write the equations for the oxidation to their oxides of

a magnesium, b lithium, c aluminium,
d carbon, e hydrogen.

See if you can work out the formulae of the oxides without looking at the line below. See Chapter 3.

Formulae [Li_2O, CO_2, Al_2O_3, MgO, H_2O]

Reduction

This, at its simplest, means removal of oxygen.

Hot copper oxide can be reduced to copper using hydrogen gas.

$$CuO + H_2 \rightarrow Cu + H_2O$$

Notice that the hydrogen has been oxidised in this reaction.

Redox reactions

The definition of oxidation and reduction has been extended to include other reactions which may not include oxygen at all!

Look at the reaction below.

$$2\,Cu + O_2 \rightarrow 2\,CuO$$

The copper is neutral at the beginning of the reaction, but has formed a Cu^{2+} ion when it reacts with oxygen.
It has lost electrons.

Oxygen is neutral at the start of the reaction and it forms an oxide ion O^{2-} when it reacts with copper.
It has gained electrons.

The extended definition of oxidation reduction reactions is:
Oxidation **i**s **l**oss of electrons
Reduction **i**s **g**ain of electrons (Use the phrase OIL RIG to recall this)
and the general term is **redox reactions.**

Copper has been oxidised and oxygen reduced in the above example.
It is always the case that whenever an element is oxidised another is reduced in the same reaction.

$$2Na + Cl_2 \rightarrow 2\,NaCl$$

7 What is happening to sodium and chlorine in this reaction and why is it is a redox reaction?

Neutralisation

Another group of reactions is that of the **acids**. Acids always form **salts** when they react , so you need to know the names of the salts that are produced by each acid. Here are some common ones.

Acid	Formula	Type of salt	Example of salt
sulphuric	H_2SO_4	sulphate	Na_2SO_4
hydrochloric	HCl	chloride	KCl
nitric	HNO_3	nitrate	$Pb(NO_3)_2$
ethanoic	CH_3COOH	ethanoate	$LiCH_3COO$

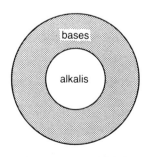

4.4 An alkali is a soluble base.

A base is a metal oxide, such as copper oxide, or hydroxide, such as sodium hydroxide, (although ammonium hydroxide, NH_4OH is included because the ammonium ion NH_4^+ behaves like a metal ion). Bases will neutralise acids. A base that dissolves in water, such as potassium hydroxide, is an alkali.

The main reactions are:

acid	+	reactive metal	$\rightarrow$	salt	+	hydrogen
acid	+	base*	$\rightarrow$	salt	+	water
acid	+	metal carbonate	$\rightarrow$	salt	+	water + carbon dioxide

* See margin note on bases

Ionic equations

When an acid reacts with an alkali in solution, we can write the equation in terms of ions. For example, hydrochloric acid reacts with sodium hydroxide to give sodium chloride and water:

$$HCl(aq) \quad + \quad NaOH(aq) \quad \rightarrow \quad NaCl(aq) \quad + \quad H_2O(l)$$

When you do this experiment, you start with two colourless solutions neither of which is neutral, and end up with a neutral colourless solution.

We can write the equation in terms of the ions in the solutions and see what is happening.

HCl(aq) exists as ions $H^+(aq)$ and $Cl^-(aq)$

NaOH(aq) exists as ions $Na^+(aq)$ and $OH^-(aq)$

One product of the reaction, sodium chloride, exists as ions $Na^+(aq)$ and $Cl^-(aq)$.

So we can write the reaction:

$$H^+(aq) + Cl^-(aq) + Na^+(aq) + OH^-(aq) \rightarrow Cl^-(aq) + Na^+(aq) + H_2O(l)$$

Overall what has happened is:

$$\mathbf{H^+(aq) \quad + \quad OH^-(aq) \quad \rightarrow \quad H_2O(l)}$$

The other ions have not changed. They are called spectator ions.

Whenever an acid reacts with an alkali , the overall reaction will be the same as the one above.

4.5 'Aren't you glad to be a spectator?'

If you can confidently write and balance the equation in question 8, then you can extend this to any Group I or Group II metal in the reacting substance.

8 a Magnesium and sulphuric acid react to give magnesium sulphate and hydrogen. Write (or say to yourself) the word equations for the reaction between each of the acids nitric, hydrochloric and ethanoic with magnesium. Then write balanced equations for the reactions. Pick out the formulae you will need: $Mg(NO_3)_2$, $MgSO_4$, $Mg(CH_3COO)_2$, $MgCl_2$.

  b Do the same thing as above for the reaction between sodium hydroxide, NaOH, and the four acids. Pick out the formulae you will need: Na_2SO_4, $NaCH_3COO$, $NaNO_3$, NaCl.

  c Write balanced symbol equations for the reactions between sodium carbonate and the four acids given in (b) and then magnesium carbonate and the four acids. The only new information you need is the formulae of sodium carbonate Na_2CO_3 and magnesium carbonate $MgCO_3$.

Double decomposition

The solutions of two different salts can react together and form an insoluble salt. If you were mixing the solutions together in a test tube you would see a solid appearing in the test tube. This is called a **precipitate**.

In the symbol equation, (aq) means that the salt is dissolved in water, and (s) would be the precipitate.

$$CuSO_4(aq) \quad + \quad Na_2CO_3(aq) \quad \rightarrow \quad CuCO_3(s) \quad + \quad Na_2SO_4(aq)$$

The ions have 'swapped partners'. This is called **double decomposition**.

9 Describe what you would see happening in the above reaction.

10 Sodium and potassium salts and all nitrates are always soluble in water. Use this information to write word and symbol equations for the following double decomposition reactions which take place in solution. Include state symbols in your answers.

   a lead nitrate and potassium iodide
   b potassium carbonate and barium chloride
   c sodium chloride and silver nitrate

Pick out the formulae you will need: $BaCl_2$, KI, $Pb(NO_3)_2$, K_2CO_3, $NaCl$, $AgNO_3$, $NaNO_3$, $AgCl$, PbI_2, KCl, $BaCO_3$, KNO_3

Answers

1 a magnesium + oxygen → magnesium oxide
   b (s) is solid, (l) is liquid, (g) is gas and (aq) is aqueous which means dissolved in water

2 a and d

3 a $2\,Li + F_2 \rightarrow 2\,LiF$
   b $4\,Fe + 3\,O_2 \rightarrow 2\,Fe_2O_3$

4 a $Mg \qquad + \quad 2\,HCl \quad \rightarrow \quad MgCl_2 \qquad + \quad H_2$
   b $4Na \qquad + \quad O_2 \qquad \rightarrow \quad 2\,Na_2O$
   c $Ca(OH)_2 \quad + \quad 2\,HNO_3 \rightarrow \quad Ca(NO_3)_2 \quad + \quad 2H_2O$
   d $Ca \qquad + \quad 2\,H_2O \quad \rightarrow \quad Ca(OH)_2 \quad + \quad H_2$

5 a iron + sulphur → iron sulphide
      $Fe + S \rightarrow FeS$
   b magnesium + nitrogen → magnesium nitride
      $3Mg + N_2 \rightarrow Mg_3N_2$
   c sodium + bromine → sodium bromide
      $2\,Na + Br_2 \rightarrow 2\,NaBr$

6 a $2\,Mg + O_2 \rightarrow 2\,MgO$
   b $4\,Li + O_2 \rightarrow 2\,Li_2O$
   c $4\,Al + 3\,O_2 \rightarrow 2\,Al_2O_3$
   d $C + O_2 \rightarrow CO_2$
   e $2\,H_2 + O_2 \rightarrow 2\,H_2O$

7 $Na \rightarrow Na^+$, so it is oxidised (loss of an electron); $Cl \rightarrow Cl^-$, so it is reduced (gain of an electron).

8 a $Mg + 2\,HCl \rightarrow MgCl_2 + H_2$
      $Mg + 2\,HNO_3 \rightarrow Mg(NO_3)_2 + H_2$
      $Mg + H_2SO_4 \rightarrow MgSO_4 + H_2$
      $Mg + 2\,CH_3COOH \rightarrow Mg(CH_3COO)_2 + H_2$

   b $NaOH + HCl \rightarrow NaCl + H_2O$
      $NaOH + HNO_3 \rightarrow NaNO_3 + H_2O$
      $2\,NaOH + H_2SO_4 \rightarrow Na_2SO_4 + 2\,H_2O$
      $NaOH + CH_3COOH \rightarrow NaCH_3COO + H_2O$

   c $Na_2CO_3 + 2\,HCl \rightarrow 2\,NaCl + CO_2 + H_2O$
      $Na_2CO_3 + 2\,HNO_3 \rightarrow 2\,NaNO_3 + CO_2 + H_2O$
      $Na_2CO_3 + H_2SO_4 \rightarrow Na_2SO_4 + CO_2 + H_2O$
      $Na_2CO_3 + 2\,CH_3COOH \rightarrow 2\,NaCH_3COO + CO_2 + H_2O$
      $MgCO_3 + 2\,HCl \rightarrow MgCl_2 + CO_2 + H_2O$
      $MgCO_3 + 2\,HNO_3 \rightarrow Mg(NO_3)_2 + CO_2 + H_2O$
      $MgCO_3 + 2\,H_2SO_4 \rightarrow MgSO_4 + CO_2 + H_2O$
      $MgCO_3 + 2\,CH_3COOH \rightarrow Mg(CH_3COO)_2 + CO_2 + H_2O$

9 A (blue) solution is added to a colourless solution and a (green) solid forms.

10 a $Pb(NO_3)_2(aq) + 2\,KI(aq) \rightarrow 2\,KNO_3(aq) + PbI_2(s)$
    b $K_2CO_3(aq) + BaCl_2(aq) \rightarrow 2\,KCl(aq) + BaCO_3(s)$
    c $NaCl(aq) + AgNO_3(aq) \rightarrow NaNO_3(aq) + AgCl(s)$

Chapter 5 Chemical patterns

Starting points

- Do you understand the following terms: **element, compound, relative atomic mass**? Write a sentence or two (no more) to explain what you understand by each term. Then check the glossary at the back of the book.

The Periodic Table

This is a chemical map showing all the elements in order of their **atomic number**, Z. It was first drawn up by the Russian **Dimitri Mendeleev** in 1869.

The work of Mendeleev

Mendeleev thought that if his table of the elements was to be really useful, it should be able to *predict* new information, not just explain what was already known. His original arrangement is shown below.

Mendeleev arranged all 63 elements then known in rows and columns in order of their relative atomic masses. However he realised that there might well be undiscovered elements and left gaps for them in sensible places. He predicted the properties of these unknown elements by averaging the properties of the elements on either side of the gap. The elements from the gaps have now all been discovered and fit Mendeleev's predictions extremely closely.

5.1 Mendeleev's Periodic Table.

H							
Li	Be	B	C	N	O		
Na	Mg	Al	Si	P	S	Cl	
K Cu	Ca Zn		Ti	V As	Cr Se	Mn Br	Fe Co Ni
Rb Ag	Sr Cd	Y In	Zr Sn	Nb Sb	Mo Te	I	Ru Rh Pd
Cs Au	Ba Hg	La Tl	Pb	Ta Bi	W		Os Ir Pt
I	II	III	IV	V	VI	VII	VIII

Alternative versions of the Periodic Table

There are various other possible arrangements as well as the familiar modern one shown on page 62.

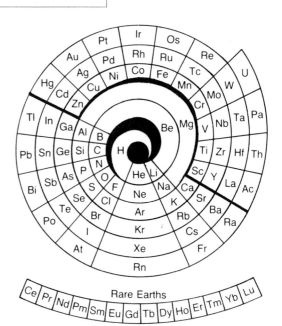

5.2 and 5.3 Two unusual forms of the Periodic Table.

5.4 The 'staircase line' divides metals (on the left) from non-metals (on the right).

Using the Periodic Table

The Periodic Table can tell us a great deal of information about the elements once we know how to use it.

Metals and non-metals

The 'staircase line' divides the metals (on the left) from non-metals (on the right). See Fig 5.4.

What properties would you expect a) the metals and b) the non-metals to have? Look back at chapter 2 if you are not sure.

Some elements near the line appear to have a combination of metallic and non-metallic properties. For example silicon, Si. This element is just on the non-metal side of the line yet it looks quite shiny, rather like a metal, and it conducts electricity to some extent so that it is called a semi-conductor. Its electrical conduction can be altered, which has led to the development of silicon chips, micro-electronics and lap-top computers. Elements like silicon are called **metalloids** or semi-metals.

1 You may never have heard of any of the following elements. Pick from them two which are definitely metals, two non-metals and one which is probably a semi-metal:

Ge, Tl, Xe, Sr, F

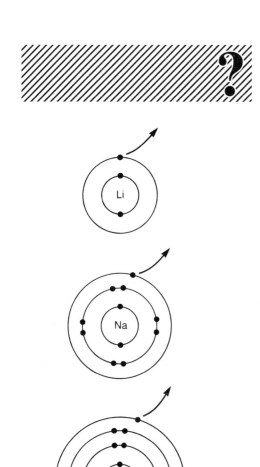

5.5 Group I atoms all lose the outer electron when they react with other atoms.

Groups

These are vertical columns in the Periodic Table. They are usually numbered from left to right missing out the central block of **transition elements**. If you look at the electron arrangements of all the elements within a particular group it is obvious that what they have in common is the same number of outer electrons.

Group I	**Group II**	**Group III**	**Group 0**
			He 2
Li 2,1	Be 2,2	B 2,3	Ne 2,8
Na 2,8,1	Mg 2,8,2	Al 2,8,3	Ar 2,8,8
K 2,8,8,1	Ca 2,8,8,2		

This means that elements in the same group will react in similar ways. They behave like a like a chemical family.

Group I the alkali metals

Each element reacts by losing its outermost electron, forming a 1+ **ion.** Losing this outer electron takes relatively little energy (the amount is called the **ionisation energy**) and so this group of elements is very reactive.

The elements react with oxygen at room temperature and therefore have to be stored out of contact with the air.

lithium   +   oxygen   →   lithium oxide
4 Li(s)   +   O₂(g)   →   2 Li₂O(s)

$$4\,\text{Li(s)} + \text{O}_2\text{(g)} \rightarrow 2\,\text{Li}_2\text{O(s)}$$

2 Predict the reaction of sodium with oxygen. Give name and formula of the product and write a balanced symbol equation for the reaction. Remember members of the same group react similarly.

The Group I elements also react with water, producing hydrogen and forming alkaline solutions of their hydroxides. This is the reason for the group's name.

lithium   +   water   →   lithium hydroxide   +   hydrogen
2 Li(s)   +   2 H₂O(l)   →   2 LiOH(aq)   +   H₂(g)

$$2\,\text{Li(s)} + 2\,\text{H}_2\text{O(l)} \rightarrow 2\,\text{LiOH(aq)} + \text{H}_2\text{(g)}$$

Note: lithium hydroxide is the alkali.

3 Write word and symbol equations for the reactions of a sodium and b potassium with water.

In all its reactions lithium is the least reactive of the alkali metals, sodium is more reactive, and potassium is even more so. We call this a **reactivity trend**. Trends allow us to make predictions.

Reactions of the alkali metals with water

Lithium   The metal floats on a bed of steam produced by the heat of the reaction.

Sodium   The reaction produces so much heat that the sodium melts and forms a molten sphere.

Potassium   Not only does the potassium melt but it ignites the hydrogen which is produced and the reaction ends with a small explosion.

5.6   Can you identify the metals reacting with water?

The reactivity trend is clear.

least reactive   →   most reactive

Li   Na   K

The atomic structures of lithium, sodium and potassium can help to explain the trend in reactivity.

Each atom reacts by losing one electron, but the electron in lithium is the closest to the pull of the positive nucleus. This makes it easier to detach the outermost electron from sodium than from lithium. This will be easier still in the case of potassium.

4 Use the information above to predict whether potassium or caesium is the more reactive. Explain your reasoning.

As well as trends in chemical properties, Group I also has trends in physical properties. Some of these are shown in Table 5.1.

Table 5.1   Properties of the alkali metals

Element	Atomic number	Electron arrangement	Density g/cm³	Melting point °C	Boiling point °C
Li	3	2,1	0.53	181	1342
Na	11	2,8,1	0.97	98	883
K	19	2,8,8,1	0.86	63	760
Rb	37	2,8,18,8,1	1.53	39	686
Cs	55	2,8,18,18,8,1	1.88	29	669
Fr	87	2,8,18,18,32,8,1	Can you predict?		

5 The properties of francium are not accurately known because it is so rare. Try to predict its density, melting point and boiling point from the trends in Table 5.1. Hint: one way to approach this is to plot graphs of property against atomic number.

Group II   the alkaline earth metals

These metals react by losing two electrons from their outer shells, forming 2+ **ions**. This requires more energy than just losing one electron so these elements are less reactive than the corresponding elements in Group I (but more reactive than those in Group III).

Their chemical properties are broadly similar to those of the Group I elements.

6 What do you expect the reactivity trend to be as you go down Group II from beryllium to radium ? Explain your answer.

Group VII the halogens

The halogens are a group of non-metals. The elements are similar to one another in several ways:

• they all exist as diatomic molecules, F_2, Cl_2, etc.

• they are all bleaches and have a 'bleachy' smell.

• they form compounds with similar formulae.

Again there are trends in their properties, some of which are shown in Table 5.2.

Table 5.2   properties of the halogens

Element	Atomic number	Electron arrangement	Appearance	Melting point °C	Boiling point °C
F	9	2,7	Colourless gas	−220	−188
Cl	17	2,8,7	Greenish gas	−101	−35
Br	35	2,8,18,7	Brown liquid	−7	59
I	53	2,8,18,18,7	Black solid	114	184
At	85	2,8,18,32,18,7	Can you predict?		

7 Like francium, astatine is a rare and radioactive element whose properties are not accurately known. Use the trends in Table 5.2 to help you predict its missing properties. Hint: one method is to plot a graph of the property against atomic number.

Reactions of the halogens

This time the chemical reactivity is based on getting an extra electron to fill the single gap in the outer shell.

$$F_2 \quad + \quad 2\,e^- \quad \rightarrow \quad 2\,F^-$$
fluorine   fluoride
molecule   ions

The reactivity trend in the halogens is as follows.

least reactive   →   most reactive
      At   F

This is because the pull of the nucleus is what attracts and holds the extra electron. So the smallest atom, fluorine, has the greatest pull and is the most reactive. Astatine, the largest atom, has the least pull and is least reactive.

5.7   Group VII atoms gain an electron when they react.

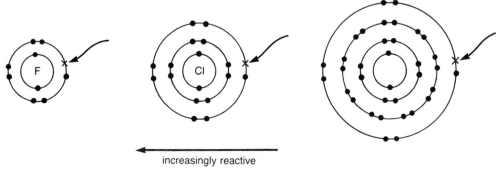

increasingly reactive

5.8   The reactivity of the elements increases as we go up the group.

The halogens are also able to fill the single gap in their outermost shell by bonding covalently. Group VII atoms can use their unpaired electron to form a single bond with another non-metal atom (see Chapter 2).

Group 0   the inert gases

This group is numbered 0, rather than VIII, because the whole group of elements was discovered after Mendeleev's first version of the Periodic Table and so it was 'tacked on' later. The whole group consists of very unreactive gases. Their unreactivity is due to the fact that they all have complete outer electron shells. They do in fact form a few compounds but the first was discovered only as recently as 1962.

Periods

These are horizontal rows in the Periodic Table.

Period 1 contains hydrogen and helium
Period 2 contains lithium, beryllium, boron, carbon, nitrogen, oxygen, fluorine and neon
and so on

As we go from left to right across a period, the outer shell of electrons is filling up, see Chapter 2.

Periods have patterns too, see Table 5.3.

Table 5.3   Trends across Period 3

Group	I	II	III	IV	V	VI	VII	O
Element	Na	Mg	Al	Si	P	S	Cl	Ar
		Metals		Semi-metal		Non-metals		(Inert gas)
		less reactive →				← *less reactive*		
Structure of element		Giant metallic		Giant covalent		Molecular		Atomic
Ion	Na^+	Mg^{2+}	Al^{3+}	None	P^{3-}	S^{2-}	Cl^-	None
Oxide	Na_2O	MgO	Al_2O_3	SiO_2	P_4O_{10} (P_4O_6)	SO_2 (SO_3)	Cl_2O (Cl_2O_7 etc.)	None
	Strongly basic		Some acidic and some basic properties		Acidic			
Structure of oxide		Giant ionic		Giant covalent		Molecular		None

As we go from left to right across a period we move from

- metals to semi-metals to non-metals.
- giant structures to molecular structures.
- basic oxides to acidic oxides.
- elements which form positive ions to elements which form negative ions.

The transition metals

The transition elements form a block, not a group or a period. In it are contained most of the everyday metals such as iron and copper. These elements have typical metallic properties (see Chapter 2). They are much less reactive than the metals in Groups I and II which makes them more useful – you could hardly make a car out of sodium! They often form coloured compounds – a property which is caused by their incomplete inner electron shells (these are the 'reserve shells' mentioned in Chapter 2).

Answers

1 Metals: Sr and Tl   Non-metals: F, Xe   Semi-metal: Ge

2 Na_2O Sodium oxide (look at the diagrams in Chapter 2 if you had difficulty).
   $$4\,Na(s) \quad + \quad O_2(g) \quad \rightarrow \quad 2\,Na_2O(s)$$
   If you had problems, the chapter on Equations will help.

3 a sodium   + water $\rightarrow$ sodium hydroxide   + hydrogen
      $2\,Na(s)$ + $2\,H_2O\,(l)$ $\rightarrow$ $2\,NaOH(aq)$ + $H_2(g)$

   b potassium   + water $\rightarrow$ potassium hydroxide + hydrogen
      $2\,K(s)$ + $2\,H_2O(l)$ $\rightarrow$ $2\,KOH(aq)$ + $H_2(g)$

4 The outermost electron in caesium is even further away from the pull of the nucleus than that in potassium, so caesium is the more reactive.

5 Very roughly: density 2.3 g/cm^3, melting point 17 °C, boiling point 650 °C.

6 The reactivity will increase as we go from beryllium to radium as the outer electrons get further from the nucleus and are lost more easily.

7 Probably a black solid. Very roughly: melting temperature 300 °C, boiling temperature 340 °C.

8 Less reactive as they must gain two electrons rather than one for the halogen.

9 Hydrogen has one electron in its outer shell so it could react by losing one electron like Group I elements or by gaining one like Group VII elements.

Chapter 6   Chemical calculations

Starting points

- Do you understand the following terms:
  element, compound, atom, molecule, ion, giant structure? Write a sentence or two (no more) to explain what you understand by each term. Then check the glossary at the back of the book.

The masses of atoms

One of the most important things that chemists did in the last century (and it was a painstaking task) was to find out the *relative* masses of the atoms of the elements. That is, they did not know the actual mass of any atom, but they found out that an oxygen atom was 16 times the mass of a hydrogen atom, while a sulphur atom was twice the mass of an oxygen atom and so on. In this way they formed a scale of mass, with hydrogen as the lightest atom and all the other elements on the scale compared with hydrogen. These masses of the atoms of elements are called their **relative atomic masses** and the symbol for them is A_r. We can look them up in the Periodic Table as shown below.

The Periodic Table in this book has most relative atomic masses rounded to the nearest whole number. Other versions of the Periodic Table may have the relative atomic mass and the atomic number reversed but this should not put you off: the relative atomic mass is always the larger.

6.1   What combinations of atoms would balance this 'atomic see-saw'?

---

1   Use the Periodic Table to look up the relative atomic masses of the following atoms:
    a   sodium, Na   c   lead, Pb
    b   oxygen, O   d   chlorine, Cl

2   Imagine you had an atomic see-saw. A magnesium atom is placed on one side. What combinations of other atoms could be placed on the other side to balance it? For example two carbon atoms would work. There are lots of combinations. Give at least six.

Atomic masses in u

It is now possible to weigh (measure the masses of) individual atoms using an instrument called a mass spectrometer. The mass of the lightest atom, hydrogen, is 1.6×10^{-27} kg. For simplicity chemists call this mass 1 **atomic mass unit** (symbol u). So the relative atomic mass, A_r, of a hydrogen atom is 1 and its mass is 1 u. Either way the *numbers* are the same. We can, of course, do this for all the elements. A helium atom has four times the mass of a hydrogen atom, so helium's relative atomic mass is 4 and a helium atom has a mass of 4 u. A_r for lithium is 7, and a lithium atom has a mass of 7 u.

The mass spectrometer

How can we weigh atoms? While the details of the mass spectrometer are complicated, the principles are fairly straightforward. Atoms are first bombarded with a beam of electrons which knocks out an electron from each atom to make it into a positive ion. These positive ions are then attracted by a negatively charged plate. This accelerates them to a high speed. Exactly how fast depends on the mass of the atom. The heavier the atom the slower it goes. The speeding ions pass through a hole in the plate and into a magnetic field. A magnetic force then deflects the ions from their straight line path. Again the amount of deflection depends on the mass of the atom. The ions then hit a detector. We can work back from the position where the ions hit the detector to find the mass of the original atom.

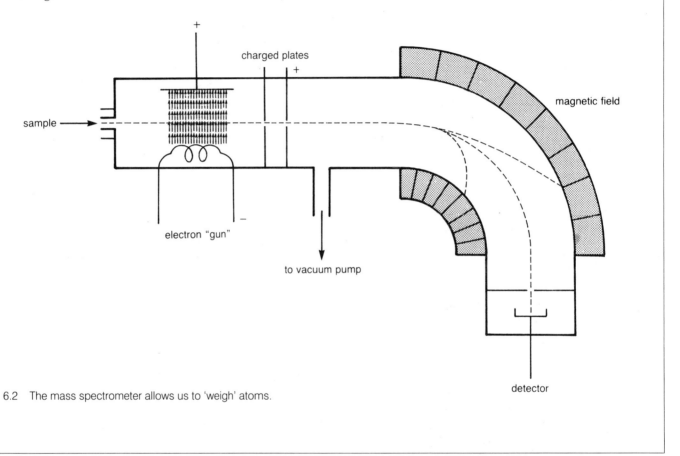

6.2   The mass spectrometer allows us to 'weigh' atoms.

Masses of molecules

The mass of a molecule is found by adding together the masses of the atoms it contains. We can either work in terms of **atomic masses** (in u) or **relative atomic masses** (no units). It is called the **molecular mass** (in u), or **relative molecular mass** (no units) which has the symbol M_r.

(i) hydrogen fluoride, HF

Atomic masses H = 1 u, F = 19 u
Molecular mass 1 u + 19 u = 20 u
Or relative molecular mass, M_r = 20

(ii) water, H_2O

Atomic masses H = 1 u, O = 16 u
Molecular mass = $(2 \times 1) + (1 \times 16) = 18$ u
Or relative molecular mass, $M_r = 18$

(iii) urea $CO(NH_2)_2$

Atomic masses C = 12 u, H = 1 u, N = 14 u, O = 16 u
Molecular mass = $12 + 16 + (14 + [2 \times 1]) \times 2 = 60$ u
Or relative molecular mass, $M_r = 60$

For compounds which do not exist as molecules such as sodium chloride (NaCl) which has a giant structure of ions (see Chapter 2) the term **molecular mass** or **relative molecular mass** is still used even though we are not talking about molecules. If we use the same term it makes everything simpler.

NaCl   23 + 35.5 = 58.5
$CaCO_3$ 40 + 12 + (3×16) = 100
$Cu(NO_3)_2$ 64 + $(14 + [3 \times 16]) \times 2 = 188$

3 Work out (using relative atomic masses in the Periodic Table) the relative molecular mass of:
   a   ammonia, NH_3
   b   calcium hydroxide, $Ca(OH)_2$
   c   oxygen molecule, O_2
   d   ethanol, C_2H_6O

Using chemical equations

We can use a **balanced equation** (see Chapter 4) to work out how much (by mass) of different substances react together. For example methane (natural gas) burns according to the following equation:

CH_4	+	$2 O_2$	→	$2 H_2O$	+	CO_2
1 particle		2 particles		2 particles		1 particle
16 u		2×32 u		2×18 u		44 u
16 u		64 u		36 u		44 u

Notice that the total mass on the left of the equation is the same as the total mass on the right (80 u). It does not matter what mass units we use. So 16 u of methane will react with 64 u of oxygen to give 36 u of water and 44 u of carbon dioxide; 16 tonnes of methane will react with 64 tonnes of oxygen to give 36 tonnes of water and 44 tonnes of carbon dioxide and so on.

6.3   In a bunsen flame, methane is combining with oxygen to produce carbon dioxide and water.

4 Work out the reacting masses in grams for:

   a $2 H_2$ + O_2 → $2 H_2O$
   b H_2 + Cl_2 → $2 HCl$

The mole

The examples above show us that substances react together in rather complicated proportions by mass – in the methane example above, 16 : 64 : 36 : 44. To make things simpler, we use a quantity called a **mole** which is the atomic or molecular mass of a substance in grams.

Using the idea that a mole of substance is its molecular mass in grams, find the mass of a mole of water, H_2O.

Molecular mass = 18 u, so a mole of water has a mass of 18 g.

5 Find the mass of a mole of:

    a   Sn atoms
    b NH_3 molecules
    c CO_2 molecules
    d   NaCl

How many moles are there in 4 g of hydrogen, H_2 ?

One mole has a mass of 2 g, so number of moles = 4/2 = 2 moles.

In general, the number of moles in a given mass of substance = mass in g / mass of one mole in g.

6 How many moles are there in

    a 40 g of calcium, Ca?
    b 980 g of sulphuric acid, H_2SO_4?
    c 22 g of carbon dioxide, CO_2?
    d 3.2 g of oxygen, O_2?

Moles and equations

In the burning of methane example we have 1 mole of methane reacting with 2 moles of oxygen molecules to produce 2 moles of water and 1 mole of carbon dioxide – much simpler than working in grams or u.

CH_4	+	$2\,O_2$	$\rightarrow$	$2\,H_2O$	+	CO_2
1 particle		2 particles		2 particles		1 particle
16 u		64 u		36 u		44 u
16 g		64 g		36 g		44 g
1 mole		2 moles		2 moles		1 mole

7 For the following equations, write below each substance the quantity which will react in (i) grams and (ii) moles.

    a $CuCO_3$ $\rightarrow$ CuO + CO_2
    b   Mg + 2 HCl $\rightarrow$ $MgCl_2$ + H_2
    c   HCl + NaOH $\rightarrow$ NaCl + H_2O

Substances react in simple whole number proportions of moles.

This is not a coincidence: a mole of *any* substance contains the same number of particles (atoms, molecules, ions).

If we weigh out a mole of carbon (12 g) and a mole of lead (207 g) those masses have the same number of particles in them. This is because the ratio of the masses of the moles is the same as the ratio of the masses of the atoms.

6.4 A pile of carbon and a pile of lead.

 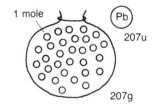

6.5 Each bag contains the same number, 6×10^{23}, of atoms and we call this amount a mole. This gives us a way of counting by weighing.

Here is another way of showing the same thing:
1 u is 1.6×10^{-24} g, so 1 mole of H atoms (1 g) contains $1/1.6 \times 10^{-24}$ atoms. This is 6×10^{23} atoms.

In the same way, 1 mole of He atoms (4 g) contains $4/6.4 \times 10^{-24}$ atoms. This is 6×10^{23} atoms again.

So 1 mole of any substance contains the same number (6×10^{23}) particles.

Since atoms and molecules react in simple whole number proportions: $1:1, 2:1, 3:1, 3:2$ etc, substances react in simple whole number proportions of moles. This is why the idea of the mole is so useful to chemists.

The formal definition of the mole is the quantity of substance which contains the same number of particles as there are atoms in 1 g of hydrogen (or strictly, 12 g of ^{12}C – see Chapter 2). However in practical terms you will find it easier to think of a mole as the atomic or molecular mass in grams or 6×10^{23} particles. The only acceptable abbreviation for mole is mol.

- Note An atom of hydrogen has a mass of
  $1 \text{ u} = 1.6 \times 10^{-24}$ g.
  An atom of helium has a mass of
  $4 \text{ u} = 4 \times 1.6 \times 10^{-24}$ g
  $= 6.4 \times 10^{-24}$ g.

Using the idea of moles

Finding formulae

18 g of water contains 2 g of hydrogen combined with 16 g of oxygen, A_r O = 16, A_r H = 1. What is its formula?

2 g of hydrogen is $2/1 = 2$ mol

16 g of oxygen is $16/16 = 1$ mol

So 2 mol of hydrogen combines with 1 mol of oxygen. Since a mole of all substances contains the same number of particles, 1 *atom* of oxygen combines with 2 *atoms* of hydrogen: the formula must be H_2O.

8 What is the formula of the following compounds?

a 196 g of sulphuric acid contains 4 g of hydrogen, 64 g of sulphur and 128 g of oxygen.
b 12.4 g sodium oxide contains 9.2 g of sodium and 3.2 g of oxygen.
c 17 g of ammonia contains 14 g of nitrogen and 3 g of hydrogen.
d 6.3 g of nitric acid contains 0.1 g of hydrogen, 1.4 g of nitrogen and 4.8 g of oxygen.

Concentration

We use the mole to define **concentration** – how much of a substance (called the solute) is dissolved in water (or other solvent) when we make a solution.

Concentrations are measured in moles per litre (mol/L)

Definition

A 1 mol/L solution contains 1 mole of solute per litre (1000 cm³). So a 2 mol/L solution contains 2 moles of solute per litre and so on.

9 What is the concentration in mol/L of
a 2 moles of H_2SO_4 in 1000 cm³ of solution?
b 2 moles of H_2SO_4 in 2000 cm³ of solution?
c 0.5 moles of H_2SO_4 in 100 cm³ of solution?

10 How would you make a
a 0.1 mole per litre solution of copper sulphate (relative molecular mass 160)?
b 1 mole per litre solution starting with 16 g of copper sulphate?

We may often need to know how many moles of solute are dissolved in a certain volume of solution.

For example, how many moles of silver nitrate are dissolved in 25 cm³ of 0.1 mol/L solution?

This can be worked out as follows:

1000 cm³ of solution contains 0.1 mol of solute

1 cm³ of solution contains 0.1/1000 mol of solute

25 cm³ of solution contains (0.1/1000) x 25 mol of solute

This is 2.5×10^{-3} mol.

We can use the formula:
number of moles of solute = M × V / 1000,
where M is the concentration of the solution in mol/L and V the volume of the solution in cm³.

Check that you get the same answer as above by using the formula.

Notice that the calculation does not have anything to do with the *nature* of the solute – only its concentration – so that the answer would be the same whatever the solute.

11 How many moles of solute are there in

a 10 cm³ of 1 mol/L solution?
b 50 cm³ of 2 mol/L solution?
c 20 cm³ of 0.01 mol/L solution?

Making solutions

We cannot weigh out quantities directly in moles, only in grams, so we must convert the number of moles we require into grams when making a solution.

How do we make 500 cm³ of a 0.1 mol/L solution of sodium hydroxide, NaOH?

$$
\begin{aligned}
\text{No. of moles} &= M \times V/1000 \\
&= 0.1 \times 500/1000 \\
&= 0.05 \text{ mol}
\end{aligned}
$$

The relative molecular mass of sodium hydroxide (NaOH) is 23 + 16 + 1 = 40, so 1 mole of sodium hydroxide has a mass of 40 g. So we must dissolve 0.05 × 40 = 2 g of solid sodium hydroxide and add water until the volume is 500 cm³. Note that this is not the same as dissolving 2 g of solid in 500 cm³ of water – this would give a total volume of more than 500 cm³.

When making a solution, the solute we are using *does* affect the mass of substance we must weigh out. What mass of *potassium* hydroxide would we need to weigh out to make 500 cm³ of 0.1 mol/L solution?

12 Describe how you would make the following.

a 1 dm³ of 2 mol/L sodium chloride solution (NaCl).
b 2 dm³ of 0.01 mol/L of silver nitrate solution (AgNO₃).
c 250 cm³ of 0.5 mol/L of sodium thiosulphate solution
    (Na₂S₂O₃).
d 100 cm³ of 1 mol/L iodine solution (I₂).

Using concentrations

Finding the proportions in which substances react.

100 cm³ of 1 mol/L hydrochloric acid is completely neutralised by 50cm² of 2 mol/L sodium hydroxide. In what proportion do the two react?

Using the formula no. of moles = M × V/1000
No. moles of hydrochloric acid = 1 × 100/1000 = 0.1 mol
No. moles of sodium hydroxide = 2 × 50 /1000 = 0.1 mol
So the substances react 1:1.

> 13 a 100 cm³ 1 mol/L sulphuric acid is just neutralised by
>   50 cm³ of 4 mol/L of potassium hydroxide. In what
>   proportion do they react?
>   b 20 cm³ of 0.1 mol/L of nitric acid is just neutralised by
>   10 cm³ of 0.2 mol/L sodium hydroxide. In what
>   proportions do they react?

Moles of gases

It is not easy to measure the mass of a gas but we can measure gas
volumes easily. The volume of a mole of *any* gas is about 24 000 cm³
at normal temperatures and pressures whatever the relative
molecular mass. This may seem strange but it is because the particles
in all gases are spaced out to the same extent (at the same
temperature and pressure). So to find out the approximate number of
moles in a sample of a gas at room temperature and pressure we
must divide its volume in cm³ by 24 000.

A reaction produces 120 cm³ of hydrogen gas. How many moles of
gas is this?

Moles of gas = 120/24 000 = 0.005 mol

> 14 How many moles are there in the following volumes of
>   oxygen (O_2)?
>
>   a 240 000 cm³   b 120 cm³   c 48 cm³
>   d What difference (if any) would it make if the gas were
>   neon?

Moles of liquids

Liquids can be weighed out in the same way as solids but it is often
more convenient to measure them out by volume, using a pipette or
measuring cylinder. In this case we need to use the **density** of the
liquid to calculate the volume we need. The density of a substance is
the mass in grams of 1 cm³ (or the mass in kg of 1 m³) .

$$\text{density} = \frac{\text{mass}}{\text{volume}}$$

What is the volume of 100 g of mercury, density 13.6 g/cm³?

density = mass/volume
volume = mass/density
       = 100/13.6
volume = 7.35 cm³

We need to measure out 0.1 of a mole of ethanol whose density is 0.8 g/cm³. What volume do we need? The relative molecular mass of ethanol (C_2H_6O) is 46. So the mass of 1 mole is 46 g and that of 0.1 mole is 4.6 g.

density	=	mass/volume
So volume	=	mass/density
	=	4.6/0.8
	=	5.75 cm³

15 Work out the volume of the following.

a 250 g of trichloroethane, density 1.3 g/cm³
b 1000 g of water, density 1 g/cm³
c 2 kg of ethanol, density 800 kg/m³
d 2 moles of trichloroethane, $C_2H_3Cl_3$
e 0.5 mole of water, H_2O
f 2 moles of ethanol, C_2H_6O, density 0.8 g/cm³

How to measure out quantities in moles – Summary

1 Solids
Use no. of moles = mass in grams/relative molecular mass

2 Liquids
Use no. of moles = mass in grams/relative molecular mass
and density = mass/volume

3 Solutions
Use no. of moles of solute = M × V/1000

4 Gases
At room temperature and pressure use the approximate expression
no. of moles = volume in cm³/24 000

What use are chemical calculations? (1)

Copper ores

Copper can be extracted from two ores – copper carbonate, $CuCO_3$, (malachite) or copper iron sulphide, $CuFeS_2$, (copper pyrites). Which contains the greater percentage of copper?

Malachite

The relative molecular mass of $CuCO_3$ is $64 + 12 + (3 \times 16) = 124$.
So 1 mole has a mass of 124 g.
Of this, 64 g is copper. So the % of copper is $64 \times 100/124 = 51.6\%$.

Copper pyrites

The relative molecular mass of $CuFeS_2$ is $64 + 56 + (2 \times 32) = 184$.
So 1 mole has a mass of 184 g.
Of this, 64 g is copper. So the % of copper is $64 \times 100/184 = 34.8\%$.

6.6 Quarrying for copper ore.

So malachite is the better ore in terms of % copper. However, copper pyrites also contains some iron. Calculate the % of iron.
What other factors would you need to take into account when deciding which ore to buy for processing?

What use are chemical calculations? (2)

Analysing antacids

Antacid tablets can be taken to cure acid indigestion. Some types contain magnesium hydroxide, an alkali, which reacts with hydrochloric acid, the acid present in our stomachs, according to the equation below.

$$Mg(OH)_2 \; + \; 2\,HCl \; \rightarrow \; MgCl_2 \; + \; 2\,H_2O$$

One brand of antacid is claimed to contain 0.3 g of magnesium hydroxide per tablet. One tablet sampled from the production line was found to react with 10 cm³ of 1 mol/L hydrochloric acid. Is the tablet up to specification?

No. of moles of hydrochloric acid $= M \times V/1000$
$$= 1 \times 10/1000 = 0.01 \text{ mol}$$

From the equation, each mole of hydrochloric acid reacts with $\frac{1}{2}$ mole of magnesium hydroxide. So the tablet must contain $0.01/2 = 0.005$ mol of magnesium hydroxide.

The relative molecular mass of $Mg(OH)_2$ is $24 + [2 \times (16 + 1)] = 58$.

So 1 mole of magnesium hydroxide has a mass of 58 g.

The tablet contains 0.005 mol so it contains $0.005 \times 58 = 0.29$ g.

The tablet is below specification and should be rejected.

Devise a practical method for carrying out this quality control procedure. Try to write out the method so that it could be followed by one of your classmates on work experience.

Answers

1 a 23; b 16; c 207; d 35.5

2 Any combination of atoms whose A_rs add up to 24, for example 24 hydrogens, 1 carbon plus 12 hydrogens etc.

3 a 17; b 74; c 32; d 46

4 a $4\,g\,H_2 + 32\,g\,O_2 \rightarrow 36\,g\,H_2O$
   b $2\,g\,H_2 + 71\,g\,Cl_2 \rightarrow 73\,g\,HCl$

5 a 119 g; b 17 g; c 44 g; d, 58.5 g

6 a 1; b 10; c 0.5; d 0.1

7 a
$CuCO_3$	$\rightarrow$	CuO	$+$	CO_2
124 g		80 g		44 g
1 mole		1 mole		1 mole

   b
Mg	$+$	$2\,HCl$	$\rightarrow$	$MgCl_2$	$+$	H_2
24 g		73 g		95 g		2 g
1 mole		2 moles		1 mole		1 mole

   c
HCl	$+$	$NaOH$	$\rightarrow$	$NaCl$	$+$	H_2O
36.5 g		40 g		58.5 g		18 g
1 mole		1 mole		1 mole		1 mole

8 a H_2SO_4; b Na_2O; c NH_3; d HNO_3

9 a 2 mol/L; b 1 mol/L; c 5 mol/L

10 a Dissolve 16 g of copper sulphate in 1 L of solution (or any other quantities in the same proportions).
    b 16 g of copper sulphate dissolved in 100 cm³ of solution.

11 a 0.01; b 0.1; 2×10^{-4}

12 a Dissolve 117 g of sodium chloride in 1 L of solution.
    b Dissolve 3.4 g of silver nitrate in 2 L of solution.
    c Dissolve 19.75 g of sodium thiosulphate in 250 cm³ of solution.
    d Dissolve 25.4 g of iodine in 100 cm³ of solution.

13 a 2 moles of potassium hydroxide to 1 of sulphuric acid.
    b 1 to 1.

14 a 10; b 5×10^{-3}; c 2×10^{-3}; d None

15 a 192.3 cm³; b 1000 cm³; c 0.0025 m³; d 205.4 cm³; e 9 cm³; f 115 cm³

Chapter 7   Physics background

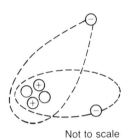

7.1   A helium atom. It is held together by the attraction of opposite electric charges. The two protons in the nucleus are held together against their electrical repulsion by an even stronger force – the nuclear force.

None of the sciences stands alone – to fully understand an advanced course in chemistry, you will need to follow some ideas which come under the heading of physics.

Electricity

Electricity and chemistry are closely linked – it is the attraction of opposite electric charges that holds electrons in atoms.

Electric charge

There are two sorts of electric charge, positive and negative. Like charges repel, so two electrons (each negative) repel and two protons (each positive) also repel. Opposite charges attract so that a proton and an electron attract. Charge is measured in **coulombs**, symbol C. An electron has a charge of 1.6×10^{-19} coulombs so it takes 6.25×10^{18} (625 followed by 16 zeros) electrons to make a coulomb of charge.

Electric circuits

The following symbols are used in electrical circuit diagrams:

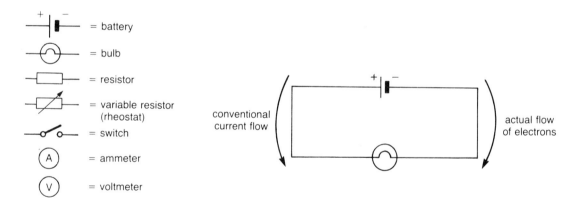

- = battery
- = bulb
- = resistor
- = variable resistor (rheostat)
- = switch
- (A) = ammeter
- (V) = voltmeter

7.2   The directions of current flows in a circuit.

In an electric circuit, charge flows through wires. The actual charge is carried by electrons. These have a negative charge and therefore flow from the negative terminal to the positive. Often we picture imaginary positive charge-carriers flowing from positive to negative. This is called the conventional current. See Fig 7.2.

Electric current

The rate at which electric charge flows through a circuit is called the **current** and is measured in amps (short for amperes). A large current could be produced by a lot of charge moving slowly or a small amount of charge moving fast. A current of one amp is a flow of charge of one coulomb per second.

Amps are coulombs per second.

Using the symbols:

Charge, **Q**, measured in coulombs
Current, **I**, measured in amps
Time, **t**, measured in seconds

I = Q/t

Ten coulombs of charge flow through an ammeter in five seconds. What current will the ammeter read?
I = Q/t
I = 10/5
I = 2 amps

A TV set draws a current of 1.6 amps for 10 000 seconds (about 3 hours). How many coulombs of charge pass through the set?
I = Q/t, so Q = I × t
Q = 1.6 × 10 000 C
Q = 16 000 C

Incidentally, since an electron has a charge of 1.6×10^{-19} C,
$16\ 000/1.6 \times 10^{-19} = 1 \times 10^{23}$ (1 followed by 23 zeros) electrons passed through the TV.

1 What is the current if
   a   four coulombs flows past a point in two seconds?
   b   100 coulombs flows past a point in 20 seconds?
   c   360 coulombs flows past a point in a minute?

2 How many coulombs of charge pass a point if
   a   a current of one amp flows for 10 seconds?
   b   a current of four amps flows for one hour?

Potential difference

Electric charge flows (called an electric current) if there is a difference in electrical **potential** between two points in a circuit. This is rather like the fact that a ball will roll if there is a height difference between two points. This **potential difference** gives energy to the electric charge. It is measured in volts and is often loosely called **voltage**.

A power supply of one volt gives a charge of one coulomb one **joule** (symbol J) of energy. The charge then transfers this energy into other forms (eg heat, light etc) in the components in the circuit. The charge is not used up, it merely travels around the circuit carrying energy. There is an analogy with postmen delivering letters. The postmen (the charge, measured in coulombs) pick up letters (the energy, measured in joules) at the post office (power supply). They deliver the letters (transfer their energy) as they go round their route and return empty-handed to the post office. The postmen are not used up!

Volts are joules per coulomb.

Using the symbols:

Potential difference, **V**, measured in volts
Charge, **Q**, measured in coulombs
Energy, **E**, measured in joules

V = E/Q

While starting a car, a 12 V battery supplies a current of 200 A for 5 seconds. How much energy has it supplied?

200 A for 5 seconds means the total charge supplied was $200 \times 5 = 1000$ C
V = E/Q, so $E = Q \times V$
$E = 1000 \times 12$
$E = 12\ 000$ J

3 A charge of 2 C is delivered by a 1.5 V battery. How much energy does the battery supply?

Batteries

Given the chance, all metal atoms try to form positive ions by giving away electrons. So in the set up in Fig 7.3, zinc atoms try to form Zn^{2+} ions.

$$Zn \rightarrow Zn^{2+} + 2e^- \text{ (equation 1)}$$

In the same way, copper atoms try to form Cu^{2+} ions.

$$Cu \rightarrow Cu^{2+} + 2e^- \text{ (equation 2)}$$

Zinc is the more reactive, so zinc atoms turn into zinc ions more readily than copper turns into copper ions and more electrons build up on the zinc terminal than on the copper one. They can then flow through the circuit to the copper. This forces copper ions to accept electrons and become copper atoms. This is the reverse of equation 2 above. The net result is that zinc dissolves, copper is deposited and electrons flow through the circuit. The energy for this comes from the chemical reactions. This is the basis of the electric cell called the Daniell cell which produces a potential difference of 1.1 V. Torch batteries work on the same principle.

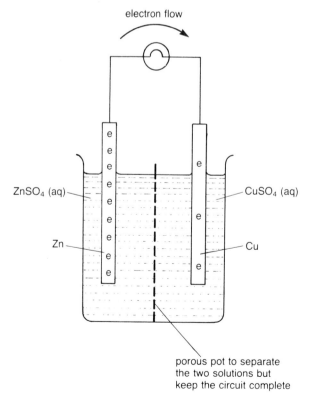

7.3 The Daniell cell.

Resistance

Every component in an electric circuit (eg bulb, motor, etc) has some **resistance** to the flow of electricity. This is measured in **ohms**, symbol, Ω). The bigger the resistance, the smaller the flow of current through the component. But, the bigger the potential difference across the component, the greater the current flow through it. The current flowing through a component can be found using the relationship which is sometimes called **Ohm's law**:

I = V/R

where
I = current in amps
V = potential difference in volts
R = resistance in Ω

In the circuit shown in Fig 7.4, what is the resistance of the component Z?

I = V/R, so R = V/I
R = 12/2 = 6 Ω

7.4

4 What current would flow in the circuit in Fig 7.5, if the values of V and R were:

   a   V = 12 V, R = 24 Ω
   b   V = 240 V, R = 480 Ω
   c   V = 1.5 V, R = 3 Ω ?

5 A torch bulb needs a current of 0.1 A and is powered by a 3 V battery. What must its resistance be?

7.5

6 What potential difference do we need to drive a current of 2 A through a resistance of 8 Ω?

Waves

Waves are regular disturbances in a **medium** like ripples in a pond. They carry energy through the medium. For example, throwing a stone into a pond produces a wave through the water which carries energy to move a toy boat up and down. See Fig 7.6.

7.6 Waves carry energy through a medium.

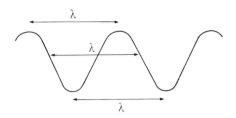

7.7 The wavelength, λ, is the distance between any successive pair of identical points on the wave.

A wave has three important properties which we can measure.

1 The **wavelength**, symbol λ (lambda). This is the distance between two successive wave crests (or troughs, or any other identical points on the wave). It is measured in metres (or cm, or other length units).

2 The **frequency**, symbol, **f** or sometimes **ν** (nu). This is the number of complete up and down vibrations the wave makes in one second. It is measured in vibrations per second (/s) sometimes called hertz (Hz).

3 The **velocity** of the wave (velocity is speed in a specified direction). This has the symbol **v** (or sometimes **c** for the speed of light and other electromagnetic radiation, such as radio wave, X-rays etc). This is the distance travelled by the wave in a given time. It is usually measured in metres per second (m/s).

These three quantities are linked by the equation

$v = f\lambda$.

The velocity of light is 300 000 000 m/s. What is the frequency of radio waves of wavelength 1500 m?

$v = f\lambda$, so $f = v/\lambda$
$f = 300\,000\,000/1500$
$f = 200\,000$ vibrations per second (or Hz)

7 A girl produces a wave in a skipping rope by flicking one end up and down twice per second. The wavelength is one metre. What is the speed of the wave along the rope?

8 The frequency of Radio 3 is about 1 200 000 Hz. The speed of all electromagnetic radiation (which includes radio) is 300 000 000 m/s. What is the wavelength of Radio 3?

Using light to probe atoms

You will have been told that electrons exist in 'shells' of different distance from the nucleus of the atom but how do we know this? A lot of our knowledge of electron arrangements comes from studying how light waves are absorbed and emitted by atoms. Light is absorbed and given out by atoms only at particular frequencies, which correspond to different colours. This is why sodium street lights are yellow and neon lights red. Each time a burst of light is absorbed by an atom, it forces an electron to jump up from one shell to the next. When the electron drops back down again, the atom gives out light. By studying the frequency of the light which atoms give out and take in, we can find out about the spacing of their electron shells.

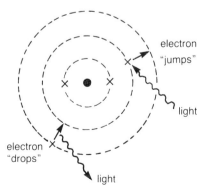

7.8 When electrons jump from shell to shell in an atom, they give out or absorb packets of energy called quanta.

7.9   Diffracted waves spread out after passing through a gap.

Interference and diffraction

These are two interesting properties of waves. **Diffraction** is shown in the photograph in Fig 7.9.

When waves pass through a small gap (about the same size as their wavelength) they spread out as shown in Fig 7.9 rather than as shown in Fig 7.10, which is what you might expect.

Interference occurs when two waves meet. If two crests meet, an extra big crest will be produced. Similarly, two troughs will produce an extra big trough. This is called **constructive interference**. However, a crest meeting a trough will cancel one another out – **destructive interference**.

7.10   If waves travelled like bullets they would not spread out after passing through a gap.

constructive interference

destructive interference

7.11   Waves adding together.

7.12

9   Try to explain the pattern in the photograph above (Fig 7.12) which shows two water waves coming together.

Answers

1   a   2 amps; b 5 amps; c 6 amps

2   a   10 C; b 14 400 C

3   3 J

4   a   0.5 amp; b 0.5 amp; c 0.5 amp

5   30 Ω

6   16 V

7   2 m/s

8   250 m

9   In some cases, the waves add together and in others they cancel out.

Chapter 8

Mathematics

You don't have to be brilliant at maths to be a good chemist, but certain mathematical skills and techniques are important.

Equations

In all sciences we look for patterns and if these are very precise we can turn them into mathematical relationships. It might be difficult to fit a mathematical expression to the connection between 'how many fillings in your teeth' and 'how many times you brush your teeth a day' although there could well be a relationship. But we can take more straightforward measurements of things that vary with each other (called **variables**) and write equations which show the connection.

If we double the temperature (in **kelvin**) of a fixed volume of gas, the pressure doubles too. Mathematically speaking, the pressure, **P**, is directly proportional to the temperature, **T**. We write:

P ∝ T

∝ means 'is proportional to'

This is shown by the data in Table 8.1.

Table 8.1

Temperature K	Pressure Pa
100	1000
150	1500
200	2000
250	2500

This also means that the pressure, P, is equal to some constant number, **k**, multiplied by the temperature.

P = kT

In this case, the constant number is 10: if we multiply the temperature in K by 10, we get the pressure in Pa (Pascals).

As the pressure on a gas, which is kept at a constant temperature, goes up, its volume goes down. More precisely, if we double the pressure we halve the volume. Mathematically speaking volume, **V**, is *inversely* proportional to pressure, **P**. Or volume is proportional to $1/P$.

$$\text{V} \propto \text{1/P}$$

or $\qquad$ V = k × 1/P

or simply $\quad$ V = k/P

This is shown by the data in Table 8.2.

Table 8.2

Pressure Pa	Volume L	1/P
1	24	1
2	12	0.5
3	8	0.33
4	6	0.25

1 pascal (Pa) is a unit of pressure of 1 newton per m²

In this case the constant number is 24. If we multiply 1/P by 24, we get the volume.

1 a What would happen to the volume of a gas if we tripled the pressure at constant temperature?
 b If two variables x and y are directly proportional to each other, what happens to one if we quadruple the other?
 c Write an expression that means x is proportional to y.

Handling equations

If you can confidently do question 2, then move to the section, **Substituting into equations**, on page 54 and try that. Otherwise go to the section below, **Rearranging equations**.

> 2 The equation that connects the pressure P, volume V and temperature T of a mole of gas is
>
> **PV = RT**
>
> where P, V and T are variables and R is a constant called the gas constant.
>
> Rearrange the equation to find
>
> a P in terms of V, R and T
> b V in terms of P, R and T
> c T in terms of P, V, and R
> d R in terms of P, V and T

Rearranging equations

If we are given the expression a = b/c and we want to find a, we substitute the values of b and c into our equation, so that if b = 10 and c = 5, then a = 10/5 = 2. But what do we do if we need to find b or c ?

We need to rearrange the equation so that b (or c) appears on its own on the left-hand side of the equation like this:

b = ?

There are many techniques for doing this. If you are not happy with the method shown below then find an algebra book and revise the method you are more familiar with.

$$a = b/c \quad \text{can be written} \quad a = \frac{b}{c} \quad \text{or} \quad a = b \div c$$

Step 1: b is divided by c so we multiply both sides by the letter which prevents b being on its own, ie **c**.

So now $\quad c \times a = \dfrac{b \times c}{c} \quad$ usually written $\quad c\,a = \dfrac{b\,c}{c}$

This is fair because we have done the same to both sides of the equation.

We can now cancel the 'c's on the right since b is being both multiplied and divided by c.

$$c \times a = \frac{b \times \cancel{c}}{\cancel{c}}$$

Which leaves $\quad c \times a = b$

or b = c × a usually written in shorthand as b = c a.

We can now rearrange this equation in the same way to find c.

Step 2: $b = c \times a$ Divide both sides by **a**.

$$\frac{b}{a} = \frac{c \times a}{a}$$

We can now cancel the 'a's on the right.

$$\frac{b}{a} = \frac{c \times \cancel{a}}{\cancel{a}}$$

So $b/a = c$ or $c = b/a$

Notice that because c started on the bottom, a two-step process was necessary. We found an expression for b first and then found one for c.

A handy rule

For simple expressions of the

form $a = \dfrac{b}{c}$, $x = \dfrac{y}{z}$ etc.

some people memorise a triangle

From this they can see directly, by covering up the one they want that:

$$a = \frac{b}{c}; \quad c = \frac{b}{a} \quad x = \frac{y}{z}; \quad z = \frac{y}{x}$$

$$b = a \times c \qquad\qquad y = x \times z$$

---

3 Try the following examples to test your understanding of the method, finding the variable in brackets in terms of the others.

   a $p = \dfrac{q}{r}$ (q)   **b** $n = m\,t$ (m)

   c $g = \dfrac{f\,e}{h}$ (h)   **d** $\dfrac{p\,r}{o\,q} = s$ (r)

Cross multiplying

A useful short cut follows from the method above for rearranging equations.

    Take the equation: $a\,b = c\,d$

Any variable can be moved to the other side of the equation, but if it was multiplying on one side it must divide on the other, and if it was dividing on one side it must multiply on the other.

So $a = \dfrac{c\,d}{b}$

     $b = \dfrac{c\,d}{a}$

If you think of any equation as having the shape below then you can change the position of the variables by following the arrows:

$$\frac{A}{C} \;{\nwarrow\!\!\!\!\nearrow \atop =} \; \frac{D}{B}$$

If one or more of A, B, C and D are not single variables then move everything.

For example $(a + b) = \dfrac{d}{c}$

the whole bracket $(a + b)$ must be moved if you change it to the other side.

$$c = \frac{d}{(a + b)}$$

A pitfall to avoid

Calculators are great but you must use them with care. Here is a common mistake to watch out for.

Imagine you are working out the expression $z = \dfrac{24}{3 \times 4}$

This means that z is 24 divided by (3×4) which is 12, so that z = 2.

But it is easy to punch into your calculator $24 \div 3 \times 4 =$ which gives the answer 32 which is incorrect.

What you must do is punch in $24 \div 3 \div 4 =$ which gives the correct answer of 2.

Alternatively you could use brackets and punch in $24 \div (3 \times 4) =$ which also gives the answer 2.

'BIDMAS'

A useful way to remember the order in which to tackle the operations in mathematical equations is summarised with the word **BIDMAS**.

Brackets
Indices
Division
Multiplication
Addition
Subtraction

Some formulae

Expressions you will regularly encounter in advanced chemistry calculations for which the practice you have just done is very useful are:

$$\text{Density} = \frac{\text{Mass}}{\text{Volume}}$$

$$\text{No. of moles} = \frac{\text{Mass}}{\text{Mass of 1 mole}}$$

No. of moles in a solution =

$$\frac{\text{Molarity} \times \text{Volume}}{1000}$$

Substituting into equations

You will often be given an expression in the form of letters and you will have to substitute the correct value for each letter before you can carry out the calculation.

You must then carry out the various mathematical operations in the correct order in your head or using a calculator.

---

4 Find the value of x in the expressions below using your calculator.

a = 0.1, b = 17, c = 14

a $X = \dfrac{2\,a\,b}{c}$ **b** $X = \dfrac{a}{b\,c}$

c $X = \dfrac{(a-b)(c-a)}{2\,b}$ **d** $X = a^2\,(b-a)$

Powers

If we multiply a number by itself we say we have squared it, so 3×3 is three squared, written 3^2.

$3 \times 3 \times 3$ is three cubed, written 3^3.

$3 \times 3 \times 3 \times 3$ is called three raised to the power 4, written 3^4.

$3 \times 3 \times 3 \times 3 \times 3$ is called three raised to the power 5 written 3^5 and so on.

Using this notation, a number raised to the power 1 is just the number itself, so that, for example, $5^1 = 5$.

Any number raised to the power 0 is 1, so that, for example, $5^0 = 1$.

Your calculator will probably have a 'squaring' button (usually marked x²). To raise a number to any other power, use the x^y button as in the following example.

To find 3⁴, punch into your calculator ③ x^y ④ ≡ and you should get the answer 81.

> 5 Find the following using your calculator.
>
>   a 8²   b 2⁸   c 3³

Numbers expressed as powers of ten

It is important that you understand the meaning of numbers expressed using powers of ten in the way shown below.

1×10^2	means	10×10	$= 100$
1×10^3	means	$10 \times 10 \times 10$	$= 1000$
1×10^4	means	$10 \times 10 \times 10 \times 10$	$= 10\,000$
2×10^4	means	$2 \times 10 \times 10 \times 10 \times 10$	$= 20\,000$

This is an important shorthand for extremely large numbers. It can be extended to small numbers too.

1×10^{-2} means $\dfrac{1}{10 \times 10}$ $= 0.01$

1×10^{-3} means $\dfrac{1}{10 \times 10 \times 10}$ $= 0.001$

1×10^{-4} means $\dfrac{1}{10 \times 10 \times 10 \times 10}$ $= 0.0001$

2×10^{-4} means $\dfrac{2}{10 \times 10 \times 10 \times 10}$ $= 0.0002$

Numbers like 1×10^3 and 1×10^{-7} are often written 10^3 and 10^{-7} respectively. You need to remember the 1 when punching them into a calculator.

As these numbers get very big or very small, the beauty of this shorthand will become apparent. You will often need to use numbers such as 10^7 or 10^{-7} so think how tedious it is to write 10 000 000 or 0.0000001 at every stage of a calculation.

> 6 Write the following numbers in two other ways as above.
>
>   a 1×10^7 b   100 000
>
>   c $3 \times 10 \times 10 \times 10 \times 10 \times 10$ d 1×10^{-7}
>
>   e   0.000 000 001   f $\dfrac{1}{10 \times 10 \times 10 \times 10 \times 10 \times 10)}$

To multiply numbers expressed like this we *add* the powers (called indices) and to divide we *subtract* them.

For example: $1 \times 10^5 \times 1 \times 10^3 = 1 \times 10^8$

We can check this by writing the numbers in full:
$100\,000 \times 1000 = 100\,000\,000$

and $1 \times 10^5 / 1 \times 10^3 = 1 \times 10^2$

Check this yourself by writing the numbers in full.

7 Calculate the following in your head.

  a $1 \times 10^4 \times 1 \times 10^4$
  b $1 \times 10^4 / 1 \times 10^3$
  c $1 \times 10^5 / 1 \times 10^7$

Very big and very small numbers

Some numbers you will often meet in chemistry are:

The Avogadro constant (the number of particles in a mole): 6×10^{23}
The mass of a hydrogen atom: 1.6×10^{-27} kg
The charge of the electron: -1.6×10^{-19} C
The speed of light: 3×10^8 m/s

Imagine writing these in full each time!

The $\boxed{\text{EXP}}$ button on your calculator

It is vital that you can use your calculator correctly and efficiently to work with numbers like this. The $\boxed{\text{EXP}}$ button stands for 'times 10 raised to the power of'.

To use your calculator to type in 1×10^3, press

$\boxed{1}$ $\boxed{\text{EXP}}$ $\boxed{3}$ and you will see displayed 1. 03 (or 1.^{03} on some calculators) Don't confuse this with 1.03.

Similarly to put 1×10^{-3} into the calculator, press

$\boxed{1}$ $\boxed{\text{EXP}}$ $\boxed{3}$ $\boxed{+/-}$ and you will see displayed 1.–03 (or 1.$^{-03}$ on some calculators)

Make sure you are familiar with the type of display your calculator gives.

When you write down an answer from a calculator display, remember to write the number in full so that a display of 1. 03 (or 1.^{03}) is written 1×10^3 and a display of 1.–03 (or 1.$^{-03}$) as 1×10^{-3}.

8 Try the following calculations, using your calculator. It is a good idea to estimate your answer to see if you are using the calculator correctly.

a $\dfrac{(1 \times 10^3) \times 14}{6}$ b $(1 \times 10^3) \times (3 \times 10^9)$

c $\dfrac{6 \times 10^{23}}{1.38 \times 10^{-23}}$

More numbers

The following are all equivalent.

1000
1×10^3 0.1×10^4
10×10^2 0.01×10^5
100×10 0.001×10^6
1000×10^0 0.0001×10^7

but the most usual way of writing this is 1×10^3. This is called **standard form**.

Using logarithms

It is probably more important to a chemist that he or she can *use* logarithms than that they fully remember the origins of this mathematical idea – but for the curious, the concept of a logarithm or **log** is briefly explained below.

The $\log_{10}$, which is read 'log to the base 10' of a number is the value of the power to which 10 is raised to give the number.

1	=	1×10^0	so	$\log_{10}1$	=	0
10	=	1×10^1	so	$\log_{10}10$	=	1
100	=	1×10^2	so	$\log_{10}100$	=	2
1000	=	1×10^3	so	$\log_{10}1000$	=	3
10000	=	1×10^4	so	$\log_{10}10000$	=	4

This should show you that using logs is a convenient way of scaling down very large numbers.

9 Work out $\log_{10}$ of 1 000 000.

The pH scale

pH	concentration of H^+ in mol/L	
"bench" hydrochloric acid → 0	1·0	$= 1 \times 10^0$
1	0·1	$= 1 \times 10^{-1}$
stomach acid → 2	0.01	$= 1 \times 10^{-2}$
cola → 3	0.001	$= 1 \times 10^{-3}$
4	0.0001	$= 1 \times 10^{-4}$
5		$= 1 \times 10^{-5}$
rain → 6		$= 1 \times 10^{-6}$
tap water → 7		$= 1 \times 10^{-7}$
8		$= 1 \times 10^{-8}$
9		$= 1 \times 10^{-9}$
some household → cleaners 10		$= 1 \times 10^{-10}$
11		$= 1 \times 10^{-11}$
12		$= 1 \times 10^{-12}$
lime water → "bench" 13		$= 1 \times 10^{-13}$
sodium hydroxide → 14		$= 1 \times 10^{-14}$

You will certainly have come across this scale of measuring acidity and alkalinity, but why does it use such odd numbers? Acidity is caused by hydrogen ions, H^+, dissolved in water. The more concentrated the H^+ ions, the more acidic the solution. The concentration of H^+ in 1 mol/L hydrochloric acid ('bench HCl') is 1 mol/L, in pure water it is 10^{-7} mol/L and in 1 mol/L sodium hydroxide ('bench NaOH') it is 10^{-14} mol/L. Yes even in an alkali, there are a few H^+ ions! These numbers cover an inconveniently large range so we get round this by using logs.

$\log_{10} 1 = 0$
$\log_{10} 1 \times 10^{-7} = -7$
$\log_{10} 1 \times 10^{-14} = -14$

This means that in most everyday solutions, the $\log_{10}$ of the H^+ concentration will be a negative number. So we agree to multiply them all by –1 to make them all positive . This is called the **pH**.

so

1 mol/L HCl has pH 0
Pure water has pH 7
1 mol/L NaOH has pH 14

8.1   The pH scale. Could you complete the table?

10 Vinegar is a solution of ethanoic acid where the concentration of H^+ is about 10^{-4} mol/L. What is its pH?

8.2   The pHs of vinegar and cola are about the same.

Logs, antilogs and your calculator

It is not too hard to work out $\log_{10}$ of numbers like 100 or 1×10^5 but we can find $\log_{10}$ of *any* number using a calculator.

To find $\log_{10} 9$ follow this procedure:

Push the buttons in the order

 and you should see display 0.954242509

---

11 Try the following examples to see that you can do this correctly. Show your answer to three decimal places.
  a $\log_{10} 14$ b $\log_{10} 7$ c $\log_{10} 9.9$
  d $\log_{10} 0.7$ e $\log_{10} 0.003$

The answers to (d) and (e) will be negative, since $\log_{10} 1 = 0$, and any number smaller than 1 has a negative $\log_{10}$.

It is also important to be able to reverse the process. Suppose you have the value of a $\log_{10}$ and you want to find the number it came from.

3.5 is the log of what number?

Push the buttons on your calculator in this order

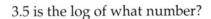

You should see display 3162.27766

This is called the **antilog** of 3.5.

so $\log_{10} 3162.27766 = 3.5$

---

12 Find the antilog of the following numbers. Show your answer to three decimal places.
  a 1.7   b 2.8   c 3.9   d 0.2
  e 0.07   f 0.0094

Using proportion

There are a lot of proportion calculations in chemistry and there are many ways of approaching them. The following method works perfectly well, but if you have a different method which works for you, then stick with it.

A 150 cm³ of a gas has a mass of 0.1 g. What is the mass of 24 000 cm³ (1 mole)?

150 cm³ of a gas has a mass of 0.1 g
1 cm³ would have a mass of 0.1/150 g (= 6.6×10^{-4} g)
24 000 cm³ would have a mass of $6.6 \times 10^{-4} \times 24\,000$ g (= 16 g)
This is the mass of one mole of the gas so its M_r must be 16.

B 5.187 g of potassium combines with 2.128 g of sulphur. How many grams of sulphur combine with 39 g (1 mole) of potassium?

5.187 g of potassium combines with 2.128 g of sulphur

1 g of potassium combines with 2.128/5.187 g of sulphur (= 0.410 g)

39 g of potassium combines with 0.410 × 39 g of sulphur (= 16g)

16 g of sulphur is 0.5 mol. So 1 mole of potassium (K) combines with 0.5 mol of sulphur (S) so the formula of the compound must be K_2S.

13 a If 0.125 g of a solid reacts with an acid to produce 10 cm³ of a gas, how much solid would produce 22 400 cm³ of gas?

b If 0.0015 g of magnesium reacts with 0.002 g of sulphur, how many grams of sulphur will 24 g of magnesium react with?

Answers

1 a It will drop to one third of its original volume.
   b It will quadruple.
   c $x \propto y$

2 a $P = RT/V$; b $V = RT/P$; c $T = PV/R$;
   d $R = PV/T$

3 a $q = pr$; b $m = n/t$; c $h = fe/g$; d $r = soq/p$

4 a 0.2428; b 4.2×10^{-4}; c –6.909; d 0.169

5 a 64; b 256; c 27

6 a 10 000 000 or $10 \times 10 \times 10 \times 10 \times 10 \times 10 \times 10$
   b 1×10^5 or $10 \times 10 \times 10 \times 10 \times 10$
   c 300 000 or 3×10^5
   d 1/10 000 000 or $1/(10 \times 10 \times 10 \times 10 \times 10 \times 10 \times 10)$
   e 1×10^{-9}
   or $1/(10 \times 10 \times 10 \times 10 \times 10 \times 10 \times 10 \times 10 \times 10)$
   f 10^{-6} or 1/ 1 000 000

7 a 1×10^8; b 10; c 1×10^{-2}

8 a 2.33×10^3; b 3×10^{12}; c 4.34×10^{46}

9 6

10 pH = 4

11 a 1.146; b 0.845; c 0.995; d –0.155; e –2.523

12 a 50.118; b 630.957; c 7943.282, d 1.58; e 1.175;
    f 1.022

13 a 280 g; b 32 g

Glossary

Atom: The smallest particle of an element.

Atomic mass: The mass of an atom measured in u. – 1 u is 1.6×10^{-27} kg.

Atomic number: The number of protons in the nucleus of an atom – also the order of an element in the Periodic Table.

Balanced equation: An equation where the total number of each type of atom is the same on each side of the arrow.

Boyle's Law: The gas law stating that the product of pressure and volume is constant.

Charles' Law: The gas law stating that volume is proportional to temperature (measured in kelvin).

Compound: A substance composed of two or more elements chemically bonded together.

Concentration: The amount of substance dissolved in a given amount of solution – usually measured in moles per litre (mol/L).

Constant Volume Law: The gas law stating that pressure is proportional to temperature (measured in kelvin).

Coulomb: The unit of electric charge.

Covalent bonding: Bonding in which the atoms are held together by sharing electrons.

Current: The rate of flow of electric charge – measured in coulombs per second, or amps.

Density: The mass of a given volume of a substance – usually measured in grams per cm^3 (g/cm^3).

Diffraction: The spreading out of waves as they pass through a gap comparable in size with their wavelength.

Double decomposition: A reaction between two soluble salts where the ions 'swap partners' and an insoluble salt is formed.

Electron: A sub-atomic particle of relative atomic mass 1/1840 and one unit of negative charge. Electrons are found orbiting the nuclei of all atoms. They are found in shells holding successively 2, 8, 8 .... electrons.

Element: One of the 109 substances that cannot be broken down chemically into anything simpler.

Energy: The ability to do work – measured in joules.

Entity: The simplest formula unit of an element or compound.

Frequency: The number of vibrations per second made by a wave – measured in vibrations per second (Hz).

Giant structure: A regular 3-D arrangement of atoms or ions in which the bonding (ionic, covalent or metallic) extends throughout the substance.

Group: A vertical column of elements in the Periodic Table in which all the elements have some similarities because they have the same number of electrons in their outer shells.

Interference: The interaction of two waves so that two crests (or two troughs) add together while a trough and a crest cancel out.

Ion: An atom (or group of atoms) having a charge (negative or positive). Ions are formed from atoms by loss or gain of electrons.

Ionic bonding: Bonding in which the atoms are held together by the transfer of electrons from metal to non-metal atoms.

Ionisation energy: The amount of energy needed to remove an electron from an atom.

Isotopes: Two or more varieties of the same element having the same number of protons but different numbers of neutrons (and therefore different relative atomic masses).

Kelvin scale: A temperature scale starting from absolute zero (-273 °C) using degrees the same size as °C. The units are written K (not °K).

Metallic bonding: Bonding in which metal atoms are held together by pooling of electrons.

Metalloid: An element which appears close to the 'staircase line' in the Periodic Table and has some metallic and some non-metallic properties. Also called a semi-metal.

Mole: The relative atomic or molecular mass of a substance in grams. One mole of any substance contains the same number (6×10^{23}) of entities.

Molecular mass: The mass of a molecule measured in u. 1 u is 1.6×10^{-27} kg.

Molecular structure: An arrangement of atoms covalently bonded together in small groups (molecules) which are independent of each other.

Molecule: A small group of atoms held together by covalent bonding.

Neutron: A sub-atomic particle of relative atomic mass 1 and no charge. Neutrons are found in the nuclei of all atoms except hydrogen.

Nucleus: The tiny central core of the atom containing protons and neutrons.

Period: A horizontal row of elements in the Periodic Table.

Potential difference: The amount of energy needed to move one coulomb of charge between two points in a circuit. More loosely the force pushing charge around the circuit. Measured in joules per coulomb or volts. Often called 'voltage'.

Precipitate: Small particles of insoluble solid which are produced in a liquid by a chemical reaction. They sink to the bottom of the liquid.

Proton: A sub-atomic particle of relative atomic mass 1 and with one unit of positive charge. Protons are found in the nuclei of all atoms.

Relative atomic mass: The mass of one atom of an element on a scale on which the mass of a hydrogen atom is one (more precisely, the mass of an atom of ^{12}C is 12).

Relative molecular mass: The mass of a molecule on the relative atomic mass scale.

Resistance: A measure of a component's opposition to the flow of electric current – measured in ohms (Ω).

Salt: An ionic compound formed by the reaction of an acid with a base.

Standard form: A way of writing a number using powers of ten so that there is one digit before the decimal point. For example 4.5×10^3.

Transition element: One of the block of typical metal elements in the centre of the Periodic Table.

Valency: The combining power of an atom.

Wavelength: The distance between two crests, troughs or any other similar points on a wave.

The Periodic Table

1 H Hydrogen 1																	2 He Helium 4
3 Li Lithium 7	4 Be Berylium 9											5 B Boron 11	6 C Carbon 12	7 N Nitrogen 14	8 O Oxygen 16	9 F Fluorine 19	10 Ne Neon 20
11 Na Sodium 23	12 Mg Magnesium 24											13 Al Aluminium 27	14 Si Silicon 28	15 P Phosporous 31	16 S Sulphur 32	17 Cl Chlorine 35.5	18 Ar Argon 40
19 K Potassium 39	20 Ca Calcium 40	21 Sc Scandium 45	22 Ti Titanium 48	23 V Vanadium 51	24 Cr Chromium 52	25 Mn Manganese 55	26 Fe Iron 56	27 Co Cobalt 59	28 Ni Nickel 59	29 Cu Copper 63.5	30 Zn Zinc 65.4	31 Ga Gallium 70	32 Ge Germanium 73	33 As Arsenic 75	34 Se Selenium 79	35 Br Bromine 80	36 Kr Krypton 84
37 Rb Rubidium 86	38 Sr Strontium 87	39 Y Yttrium 89	40 Zr Zirconium 91	41 Nb Niobium 93	42 Mo Molybdenum 96	43 Tc Technetium 99	44 Ru Ruthenium 101	45 Rh Rhodium 103	46 Pd Palladium 106	47 Ag Silver 108	48 Cd Cadmium 112	49 In Indium 115	50 Sn Tin 119	51 Sb Antimony 122	52 Te Tellurium 128	53 I Iodine 127	54 Xe Xenon 131
55 Cs Caesium 133	56 Ba Barium 137	57 ► La Lanthanum 139	72 Hf Hafnium 179	73 Ta Tantalum 181	74 W Tungsten 184	75 Re Rhenium 186	76 Os Osmium 190	77 Ir Iridium 192	78 Pt Platinum 195	79 Au Gold 197	80 Hg Mercury 201	81 Tl Thallium 204	82 Pb Lead 207	83 Bi Bismuth 209	84 Po Polonium 210	85 At Astatine 210	86 Rn Radon 222
87 Fr Francium 223	88 Ra Radium 226	89 ►► Ac Actinium 227	104 Unq Unnilquadium 261	105 Unp Unnilpentium 262	106 Unh Unnilhexium 263	107 Uns Unnilseptium 263	108 Uno Unniloctium ?	109 Une Unnilennium ?									

Key

| Atomic number |
| Symbol |
| Name |
| Relative atomic mass |

► | 58
Ce
Cerium
140 | 59
Pr
Praseodymium
141 | 60
Nd
Neodymium
144 | 61
Pm
Promethium
147 | 62
Sm
Samarium
150 | 63
Eu
Europium
152 | 64
Gd
Gadolinium
157 | 65
Tb
Terbium
159 | 66
Dy
Dysprosium
163 | 67
Ho
Holmium
165 | 68
Er
Erbium
167 | 69
Tm
Thulium
169 | 70
Yb
Ytterbium
173 | 71
Lu
Lutetium
175 |

►► | 90
Th
Thorium
232 | 91
Pa
Protactinium
231 | 92
U
Uranium
238 | 93
Np
Neptunium
237 | 94
Pu
Plutonium
242 | 95
Am
Americium
243 | 96
Cm
Curium
247 | 97
Bk
Berkelium
245 | 98
Cf
Californium
251 | 99
Es
Einsteinium
254 | 100
Fm
Fermium
253 | 101
Md
Mendelevium
256 | 102
No
Nobelium
254 | 103
Lr
Lawrencium
257 |

- **Note** In most versions of the Periodic Table, including the one above, the atomic number is written at the top of the element's box and the relative atomic mass at the bottom. However, when writing the symbols of individual elements, these are usually reversed: eg $^{23}_{11}\text{Na}$